GEOGRAPHY: THEORY IN PRACTICE
BOOK THREE

INDUSTRY

GEOGRAPHY: THEORY IN PRACTICE
BOOK THREE
INDUSTRY

RICHARD HUGGETT & IAIN MEYER

HARPER & ROW, PUBLISHERS

LONDON

Cambridge
Hagerstown
Philadelphia
New York

San Francisco
Mexico City
Sao Paulo
Sydney

1817

To Jane, Jamie, Sarah and Edward
Jacqueline and Lucinda

Copyright © 1981 Richard Huggett and Iain Meyer

All rights reserved
First published 1981
Reprinted 1984

Harper & Row Ltd
28 Tavistock Street
London WC2E 7PN

British Library Cataloguing in Publication Data

Meyer, Iain
    Industry.
    1. Industries, Location of
    2. Geography, Economic
    I. Title   II. Huggett, Richard
    338.09   HD58

ISBN 0 06 318164 9

Typeset by
STYLESET LIMITED
Salisbury · Wiltshire
Printed and bound by The Pitman Press, Bath

Contents

Preface

This book outlines and illustrates the chief theories and concepts of the geography of industry. These theories and concepts are explained concisely and, as far as possible, in jargon-free language. By means of carefully designed exercises, students should see how these theories work in practice. The exercises are, in the main, based on maps and tables from a wide range of up-to-date books and articles, or on data from official sources. Worked examples of the techniques used are included, and carefully posed questions encourage students to interpret and to criticize their results. At the end of each chapter is a list of books which might be helpful in answering the questions or in essay work. Though written as a course book, *Industry* has plenty of scope for teachers to bring in their own examples and sets of data.

Chapter One considers the problem of industrial location and in particular the role of cost and the role of revenue. These are studied using examples from a wide range of industries, including brick-making, iron and steel manufacture, and brewing. Chapter Two considers industrial and economic regions in the United Kingdom, socialist states, the Third World, and western cities. The chapter starts with a look at theories of regional development. Chapter Three examines the impact of industrial activities on the environment, topics covered including the pollution of the air, the pollution of water, and the creation of derelict land. There is a glossary of key terms.

Industry is the third book in the series *Geography: Theory in Practice*. Some of its contents are related to material covered in *Settlements* and *Agriculture*, the other two books in the series. Points of cross-reference are noted in the text.

Richard Huggett **Iain Meyer**
Macclesfield   Elstree

July 1980

CHAPTER ONE
THE LOCATION OF INDUSTRY

The Background

The location of industry presents the geographer with a fascinating puzzle. Why is there an aluminium smelter at Holyhead on Anglesey but not in London? Why should a cocoa and chocolate works be located in Birmingham but not in Aberystwyth? Why did the Vauxhall Motor Company set up a new plant at Ellesmere Port in Cheshire and not in Penzance? To answer questions of this sort we need to have some idea of what makes industry tick.

EXPENDITURE, REVENUE, AND PROFIT

Imagine that you have won a large sum of money. You think it would be fun to set up your own business. You decide to take a gamble and try your hand at brewing and selling beer on a modest scale. In other words, you will become what is known as an **entrepreneur**.

The first thing you will need to do is buy a suitable building. A pub would seem a likely spot to start as it would guarantee at least one outlet for your beer. Then you will need some brewing equipment. Next you will need raw materials from which to make beer — hops, sugar, glucose, malt, yeast, and water. You will also need supplies of containers in which to store the beer once you have made it — barrels and bottles. You will need a supply of power — electricity, gas, or oil. Your pub will doubtless be connected to one or more of these power sources but you will of course have to pay for what you use. There is, clearly, no point in making beer if you are unable to sell it. In other words, you will need a market. Your own pub will presumably generate enough demand to keep you in business for a while. Looking to the future though, you would presumably hope to expand production and capture some of the trade in nearby free houses.

Your aim will be to set the **expenditure** on materials and supplies against the **revenue** from the sale of beer. With luck, your revenue will exceed your expenditure and you will make a **profit**. If your profits keep rolling in and your capital increases you will have more scope for expansion and growth. Your expenditure can be broken down into two parts (Figure 1.1). Firstly, you will have **production** or **processing costs**. These are what it will cost you to make beer. They will include the cost of the pub, of brewing equipment, of labour (if you should take on any helpers), and of power. They will be influenced to a certain extent by your **scale of operation**: it will probably be relatively cheaper for you to produce two hundred gallons of beer a week than to produce one hundred. But it will be difficult for you to make beer at a competitive price if you make less than a hundred gallons a week. Secondly, you will have **transport costs**. These will consist of **procurement** and **distribution costs**. Distribution costs are the costs you will incur in selling and delivering your beer. Procurement costs are the costs you will incur in buying and bringing raw materials and supplies to your pub for processing.

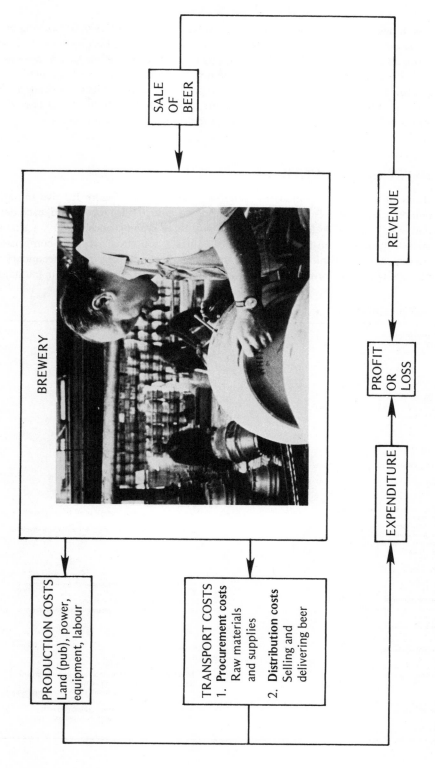

Figure 1.1 The flow of money in your brewery

Bearing all these things in mind, you will have to decide where best to set up your brewing business — you will have to decide which of the available free houses to purchase. If you could work out the likely expenditure and revenue for all possible sites, you could opt for the one which looks as if it will bring in the biggest profits. But, more likely than not, you will not have such detailed information at your disposal. The best you can do is to buy a pub which suits your needs as you see them. And if you did manage to get a business going, you might well enjoy the success of a number of very small brewing concerns which have sprung up over the last few years.

THE DIVERSITY OF INDUSTRY

Brewing is just one of a vast array of industrial activities. Table 1.1 and Figure 1.2 give an indication of how diverse industrial activity is. Notice that 27 orders of industry are recognized, each denoted by a Roman numeral. These 27 orders can be divided into three broad groups. Firstly, a group of **primary industries** concerned with the exploitation of natural resources. This group comprises agriculture, forestry, hunting, fishing, mining, and quarrying. Secondly, a group of **secondary industries** concerned with transforming natural resources. This group consists of all manufacturing

Table 1.1   British industries — a standard classification

I Agriculture, forestry, fishing
  Agriculture and horticulture
  Forestry
  Fishing

II Mining and quarrying
  Coal-mining
  Stone and slate quarrying and mining
  Chalk, clay, sand, and gravel extraction
  Petroleum and natural gas
  Other mining and quarrying

III Food, drink, and tobacco
  Grain milling
  Bread and flour confectionery
  Biscuits
  Bacon curing, meat and fish products
  Milk and milk products
  Sugar
  Cocoa, chocolate, and sugar confectionery
  Fruit and vegetable products
  Animal and poultry foods
  Vegetable and animal oils and fats
  Food industries not elsewhere specified
  Brewing and malting
  Soft drinks
  Other drink industries
  Tobacco

IV Coal and petroleum products
  Coke ovens and manufactured fuel
  Mineral oil refining
  Lubricating oils and greases

V Chemicals and allied industries
  General chemicals
  Pharmaceutical chemicals and preparations
  Toilet preparations
  Paint
  Soap and detergents
  Synthetic resins and plastics materials and synthetic rubber
  Dyestuffs and pigments
  Fertilizers
  Other chemical industries

VI Metal manufacture
  Iron and steel (general)
  Steel tubes
  Iron castings, etc.
  Aluminium and aluminium alloys
  Copper, brass, and other copper alloys
  Other base metals

VII Mechanical engineering
  Agricultururral machinery (except tractors)
  Metal-working machine tools
  Pumps, valves, and compressors
  Industrial engines
  Textile machinery and accessories
  Construction and earth-moving equipment
  Mechanical handling equipment
  Office machinery
  Other machinery
  Industrial (including process) plant and steelwork
  Ordnance and small arms
  Other mechanical engineering not elsewhere specified

VIII Instrument engineering
  Photographic and document-copying equipment
  Watches and clocks
  Surgical instruments and appliances
  Scientific and industrial instruments and systems

IX Electrical engineering
  Electrical machinery
  Insulated wires and cables
  Telegraph and telephone apparatus and equipment
  Radio and electronic components
  Broadcast receiving and sound reproducing equipment
  Electronic computers
  Radio, radar, and electronic capital goods
  Electric appliances primarily for domestic use
  Other electrical goods

Table 1.1 (continued)

X Shipbuilding and marine engineering

XI Vehicles
  Wheeled tractor manufacturing
  Motor vehicle manufacturing
  Motor cycle, tricycle, and pedal cycle manufacturing
  Aerospace equipment manufacturing and repairing
  Locomotives and railway track equipment
  Railway carriages and wagons and trams

XII Metal goods not elsewhere specified
  Engineers' small tools and gauges
  Hand tools and implements
  Cutlery, spoons, forks, and plated tableware, etc.
  Bolts, nuts, screws, rivets, etc.
  Wire and wire manufactures
  Cans and metal boxes
  Jewellery and precious metals
  Metal industries not elsewhere specified

XIII Textiles
  Production of man-made fibres
  Spinning and doubling on the cotton and flax systems
  Weaving of cotton, linen, and man-made fibres
  Woollen and worsted
  Jute
  Rope, twine, and net
  Hosiery and other knitted goods
  Lace
  Carpets
  Narrow fabrics (not more than 30cm wide)
  Made-up textiles
  Textile finishing
  Other textile industries

XIV Leather, leather goods and fur
  Leather (tanning and dressing) and fellmongery
  Leather goods
  Fur

XV Clothing and footwear
  Weatherproof outerwear
  Men's and boys' tailored outerwear
  Women's and girls' tailored outerwear
  Overalls and men's shirts, underwear, etc.
  Dresses, lingerie, infants' wear, etc.
  Hats, caps, and millinery
  Dress industries not elsewhere specified
  Footwear

XVI Bricks, pottery, glass, cement, etc.
  Bricks, fireclay and refractory goods
  Pottery
  Glass
  Cement
  Abrasives and building materials, etc. not elsewhere specified

XVII Timber, furniture, etc.
  Timber
  Furniture and upholstery
  Bedding, etc.
  Shop and office fitting
  Wooden containers and baskets
  Miscellaneous wood and cork manufactures

XVIII Paper, printing, and publishing
  Paper and board
  Packaging products of paper, board, and associated materials
  Manufactured stationery
  Manufactures of paper and board not elsewhere specified
  Printing, publishing of newspapers
  Printing, publishing of periodicals
  Other printing, publishing, bookbinding, engraving, etc.

XIX Other manufacturing industries
  Rubber
  Linoleum, plastics floor-covering, leathercloth, etc.
  Brushes and brooms
  Toys, games, children's carriages, and sports equipment
  Miscellaneous stationers' goods
  Plastics products not elsewhere specified
  Miscellaneous manufacturing industries

XX Construction

XXI Gas, electricity, and water
  Gas
  Electricity
  Water supply

XXII Transport and communication
  Railways
  Road passenger transport
  Road haulage contracting for general hire or reward
  Other road haulage
  Sea transport
  Port and inland water transport
  Air transport
  Postal services and telecommunications
  Miscellaneous transport services and storage

XXIII Distributive trades
  Wholesale distribution of food and drink
  Wholesale distribution of petroleum products
  Other wholesale distribution
  Retail distribution of food and drink
  Other retail distribution
  Dealing in coal, oil, builders' materials, grain, and agricultural
    supplies
  Dealing in other industrial materials and machinery

XXIV Insurance, banking, finance, and business services
  Insurance
  Banking and bill discounting
  Other financial institutions
  Property owning and managing, etc.
  Advertising and market research
  Other business services
  Central offices not allocable elsewhere

XXV Professional and scientific services
  Accountancy services
  Education services
  Legal services
  Medical and dental services
  Religious organizations
  Research and development services
  Other professional and scientific services

XXVI Miscellaneous services
  Cinemas, theatres, radio, etc.
  Sport and other recreations
  Betting and gambling
  Hotels and other residential establishments
  Restaurants, cafes, snack bars
  Public houses
  Clubs
  Catering contractors
  Hairdressing and manicure
  Laundries
  Dry cleaning, job dyeing, carpet beating, etc.
  Motor repairs, distributors, garage and filling stations
  Repair of boots and shoes
  Other services

XXVII Public administration and defence
  National government service
  Local government service

Figure 1.2   Various industries

industries. Building and construction work are sometimes included in it. Thirdly, a group of **tertiary industries** which includes financial, professional, and scientific services, and all other branches of the economy not involved in the production of material goods.

There are several factors to be considered when deciding where to locate an industry (Figure 1.3). It is probably true to say that no two industries have exactly the same requirements. This is because different industries need different sorts of site, different equipment, and different buildings. They need different raw materials, different types of labour, and different sources of capital. They provide different goods or services for different markets. Even within one branch of industry requirements can differ enormously. This is chiefly because some firms may be owned privately by individuals, some may be owned privately by corporations, and some may be public companies owned by the state. The size and complexity of the firms may also vary greatly. Some may be single production units. Others may be large concerns consisting of separate units occupying different sites: one for administration, one for research, one for development, one for sales, and one for production, but all under the control of a head office. Any one firm may have one big production plant, a few medium-sized production plants, or a lot of small production plants. This depends, among other things, on what is the most economical scale of operation. There is commonly an optimum size of plant for a particular industry. A car assembly plant, for instance, might need to turn out something like 100,000 cars a year to prove competitive. In many industries, the changing economic climate has meant that it is no longer possible to run a lot of small factories serving tiny areas and make a profit. A process of 'rationalization', acting through what are called **internal economies of scale**, has taken place so that now there are a few big factories serving large areas.

It is not possible to consider the requirements of an industry in isolation — many industries are interrelated and it is difficult to consider the location of one without considering the location of others. The ties between groups of industries may lead to the clustering of allied activities in small areas. This process is called **agglomeration**. The reasons agglomeration takes place are complex. **Linkage** between allied trades, as seen in the concentration of metal industries in the Birmingham and Black Country conurbation, is certainly an important factor. So too are **external economies of scale** which enable several different firms to reduce costs by being located close to one another, usually because the firms then share the costs of such things as public utilities.

You should now see why the location of industry can be a difficult but fascinating puzzle to unravel. A number of people have tried to find an underlying logic to the puzzle. They have proposed theories of industrial location. This chapter will outline a good many of these theories and examine how they apply to the real world.

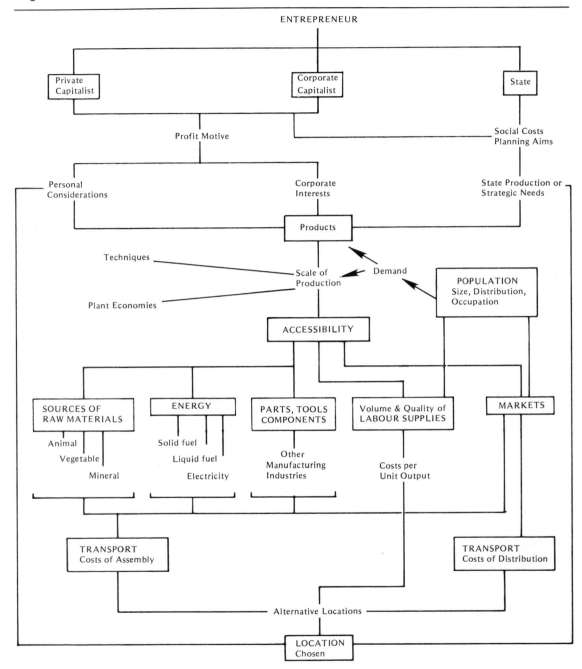

Figure 1.3   Factors an entrepreneur should consider in deciding where to set up a business
Reprinted with permission from 'Models of industrial location' by F. E. Ian Hamilton (1967) in
Models in Geography, edited by R. J. Chorley and P. Haggett, published by Methuen, London,
figure 10.1.

The Role of Cost

THE IDEAS OF ALFRED WEBER

Alfred Weber's theory of industrial location originally appeared in 1909 in a book called *On the Location of Industry*. In essence, Weber's argument is that an industry will be located where total transport costs are least; but that an industry may be attracted away from the site of lowest transport costs by sites of cheap labour or by sites where a distinct advantage accrues from agglomeration.

To reduce the complex problem of industrial location to a manageable size, Weber made the following assumptions:

i   Natural resources (raw materials) are distributed unevenly and found at just a few specific spots;

ii   the size and location of markets for the industrial products are known; and

iii   regional variations in the wage rates of a fixed labour force, always in ample supply, are also known.

Other factors which might influence the location of an industry — political and economic systems, culture, cost of land, building and equipment, interest rates, depreciation of capital, ease of transport — were all assumed to be the same throughout the region. Conditions of perfect competition were also assumed: resources and markets are of unlimited size and no factory will gain a monopoly by choosing a particular site. These assumptions can be challenged: raw materials, power, and markets are usually distributed over a large area rather than at specific spots; the framework of perfect competition lacks realism; fluctuations in demand and supply of a product lead to changes in price over time; and transport rates change with time and usually have a far more complex structure than Weber assumed. But, as with so many theories, unless gross assumptions are made, initially at least, the problem will be too complex to tackle.

LOCATING A STEEL PLANT

Weber started his analysis by looking at **transport costs**. He showed how the 'best' location for a factory can be found for the case with two sources of raw material and one market, each represented by the corners of a triangle. In Mexico, steel is made using mainly iron ore from Durango and coke from Sabinas. It is then sent to steel users in Mexico City (Figure 1.4a). Weber's analysis enables us to find the 'best' site for making steel within the **locational triangle**, as it is known, defined by Durango, Sabinas, and Mexico City. The 'best' site means the place where the costs of procuring iron from Durango and coke from Sabinas and distributing finished steel to Mexico City are cheapest.

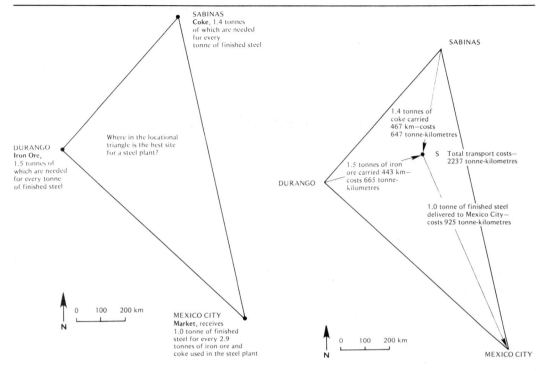

Figure 1.4a Weber's problem for the case of a steel plant in Mexico

Figure 1.4b Transport costs (weight-distance units) for a steel plant at site S

Transport costs, both procurement and distribution, can be expressed either in units of **money** or on a **weight-distance** basis. Whatever the units of measurement, all calculations must be standardized by expressing all costs relative to one tonne of product, that is, one tonne of finished steel. It takes 1.5 tonnes of iron ore and 1.4 tonnes of coke to make 1.0 tonne of finished steel and so it is the costs of transporting these amounts of materials which are computed.

The computations of transport costs, expressed in units of weight-distance, for a steel plant located at point S in Figure 1.4b are shown in Table 1.2. For every tonne of finished steel, 1.5 tonnes of iron ore are procured from Durango, 443 kilometres away. The cost of procuring iron ore for site S is thus

1.5 tonnes × 443 kilometres = 665 tonne-kilometres.

The costs of getting coke from Sabinas and delivering steel to Mexico City are calculated in like manner. They are 647 tonne-kilometres and 925 tonne-kilometres respectively. The total transport cost for a steel plant at site S is the sum of all procurement and distribution costs, that is,

665 + 647 + 925 = 2237 tonne-kilometres.

The computations of transport costs, expressed in units of money, for a steel plant located at point S in Figure 1.4b are shown in Table 1.3.

Table 1.2   Calculating transport costs on a weight-distance basis

Site and material	Weight of material to be hauled (tonnes)	Distance from steel plant (kilometres)	Transport cost (tonne-kilometres)
Durango (iron ore)	1.5	443	665
Sabinas (coke)	1.4	462	647
Mexico City (steel)	1.0	925	925
Total transport cost			2237

Table 1.3   Calculating transport costs in units of money (pesos)

Site and material	Weight of material to be hauled (tonnes)	Transport rate (pesos per kilometre)	Distance from steel plant (kilometres)	Transport costs (pesos)
Durango (iron ore)	1.5	0.050	443	22.15
Sabinas (coke)	1.4	0.075	462	34.65
Mexico City (steel)	1.0	0.050	925	46.25
Total transport cost				103.05

Notice that these calculations incorporate the actual cost of transporting raw materials and product over a given distance. In other words, transport or freight rates enter the analysis. For this reason, costings done on a money basis are usually more realistic than costings worked out on a weight-distance basis.

With this background, we may set about finding the site of cheapest transport costs for a steel plant in the locational triangle (Figure 1.4b). We shall look at three methods of doing this.

Locating the Steel Plant by Construction

This method will find the cheapest site where costs are expressed on a weight-distance basis. It is quick and easy. In essence, the site where transport costs are cheapest is fixed by finding the angles A, B, and C shown in Figure 1.5a. (The site of the factory on this figure is purely diagrammatical.) The procedure runs like this:

i   Construct a triangle the length of whose sides are proportional to the weights of material to be hauled. This is done in Figure 1.5b.

ii   Measure the angles a, b, and c in this triangle with a protractor; they are, in the example, $a = 65$ degrees, $b = 75$ degrees, and $c = 40$ degrees. (Note that the angles A and a are the angles between Durango and Mexico City; angles B and b are the angles between Mexico City and Sabinas; and angles C and c are the angles between Sabinas and Durango.)

iii   The angles A, B, and C in Figure 1.5a are related to the angles a, b, and c in the following way:

$$A = 180 - a; \quad B = 180 - b; \quad C = 180 - c.$$

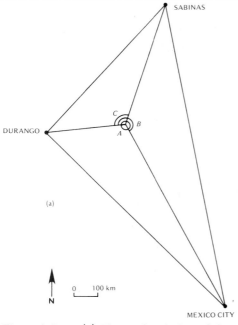

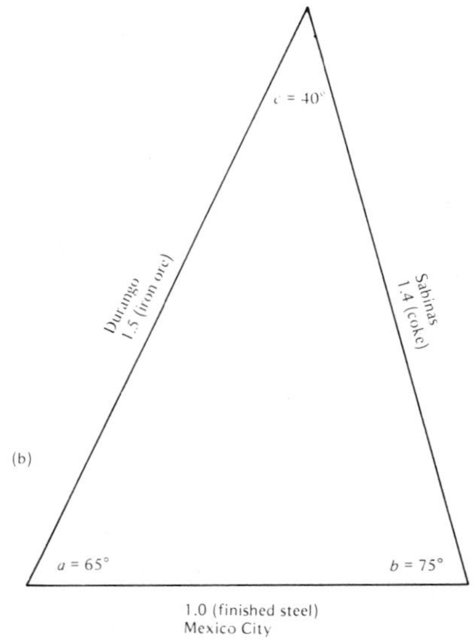

SABINAS

DURANGO

C
B
A

(a)

100 km

N

MEXICO CITY

Figure 1.5   (a) The angles A, B, and C
must be found
(b) The triangle of weights
(c) The lines which should
be drawn on tracing paper
(d) Locating the cheapest site

Figure 1.5b

c = 40°

Durango
1.5 (iron ore)

Sabinas
1.4 (coke)

(b)

a = 65°

b = 75°

1.0 (finished steel)
Mexico City

In the example we have

$A = 180 - 65 = 115$ degrees;
$B = 180 - 75 = 105$ degrees; and
$C = 180 - 40 = 140$ degrees.

iv   Plot on a piece of tracing paper three lines originating from a central point, the angles
between the lines being equal to the angles A, B, and C (Figure 1.5c).
v   Place the tracing paper with the three lines on over the map of material sources and market
(Figure 1.5d). Juggle the tracing paper until the appropriate lines pass through the material
sites and the market, that is, through Durango, Sabinas, and Mexico City. This is best done by
getting two places on their lines and then, keeping the two places on their lines, moving the
paper slowly until the third place too is on its line. When this has been done, the cheapest
site for the steel plant has been fixed — it is the point beneath the spot on the tracing paper
where the three lines originate.

1   Where would the best site be for the steel plant if the weights of materials to be hauled were iron
ore 1.6 tonnes, coke 1.4 tonnes, and finished steel 1.0 tonne?

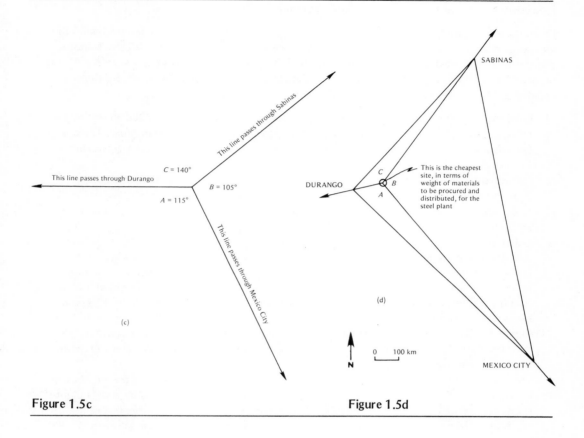

Figure 1.5c

Figure 1.5d

Locating the Steel Plant by a Mechanical Method

The best site for the steel plant, again where transport costs are expressed on a weight-distance basis, can also be found by a mechanical method devised by Varignon. Pulleys are fixed on a board at points corresponding to Durango, Sabinas, and Mexico City. Weights proportional to the weights of iron ore, coke, and finished steel are attached to three strings. (A convenient way of making these weights is to cut a length of wood or, better still, metal into sections whose lengths in, say, decimetres are equivalent to the weights of material in tonnes. In the example this would mean lengths of 1.5, 1.4, and 1.0 decimetres.)

The strings are passed over the appropriate pulleys and then tied together. One of the weights is given a little tug. The position in the locational triangle at which the knot where the three strings are tied comes to rest is the point at which the pulls of the three weights are balanced. This position corresponds to the site where transport costs for the steel plant are cheapest. It is advisable to tug the weights and then record the position of the knot at rest several times since friction between strings and pulleys may distort the results a little. Note that if the pull of any one corner of the locational triangle is greater than or equal to the sum of the pulls of the other two corners, the site with the cheapest transport costs will be the site with the overriding pull.

Locating the Steel Plant Using a Cost Surface

In a nutshell, this method involves drawing a contour map of transport costs in and around the locational triangle and then looking at the map to see where costs are lowest. It can be applied to transport costs expressed on a weight-distance basis and to transport costs expressed in units of money. An example is depicted in Figure 1.6, the construction of which used the transport rates listed in Table 1.3 and involved the following steps:

i   Concentric circles were constructed around Durango, each of which is called an **isotim** and joins sites which would incur the same costs in procuring 1.5 tonnes of iron ore. The cost at Durango is zero but increases by 0.05 pesos for every kilometre moved away from Durango (Table 1.3).

ii   Isotims were constructed around Sabinas. They join sites which incur the same costs in procuring 1.4 tonnes of coke.

iii   Isotims were constructed around Mexico City. They join sites which incur the same costs in distributing 1.0 tonne of finished steel.

iv   Using the three sets of isotims, total transport costs (procurement and distribution) were calculated and plotted on the map (Figure 1.6). This was done at all points where three isotims intersect. For example, at point A the costs are 40 pesos for procuring iron ore, 30 pesos for procuring coke, and 50 pesos for delivering finished steel. So if a steel plant were located at point A its total transport costs per tonne of finished steel would be 120 pesos.

v   Lines were drawn connecting points of equal transport costs. These are the heavy lines on Figure 1.6. They are called **isodapanes**. The isodapanes can be thought of as contours which describe the relief of the cost surface: hills are areas of high transport costs and valleys are areas of low transport costs.

vi   The position of the lowest point on the total transport cost surface was sought by studying the pattern of isodapanes. It is marked by a □. This is the best place to set up a steel plant if transport costs are to be kept as low as possible. Notice that this result differs from the solution derived earlier by construction where cost was measured in tonne-kilometres.

Steel Plants In Mexico

The steel industry in Mexico is a little more complex than has been suggested in the previous sections. Two plants, the Fundidora plant at Monterrey and the Altos Hornos plant at Monclova, produce steel (Figure 1.7a). Both plants procure iron ore from Durango and six other minor sources. Coke comes from Sabinas giving the Monclova plant a big cost advantage. On the other hand, the plant at Monterrey cuts its fuel bills by using gas which, before Mexico's own gas fields were opened, was piped from Texas. Both plants use oil from Reynosa and Tampico. Steel scrap comes in equal shares from the United States and domestic sources. Half of the domestic steel scrap comes from Mexico City and half comes from Monterrey. Finished steel from the two plants is delivered to steel-users in Mexico City and Monterrey.

R. A. Kennelly applied Weber's ideas to the location of steel plants in Mexico during the mid-1950s. To make things manageable, he assumed that iron ore came from Durango only and excluded gas from the calculations. The shaded circles in Figure 1.7a represent the weight of raw materials needed to make a tonne of finished steel. These weights are averages of data for the two plants. Kennelly assumed that three-quarters of the finished steel went to Mexico City and the remaining quarter to Monterrey. From this information, Kennelly applied several methods to find the site of cheapest transport costs in Mexico.

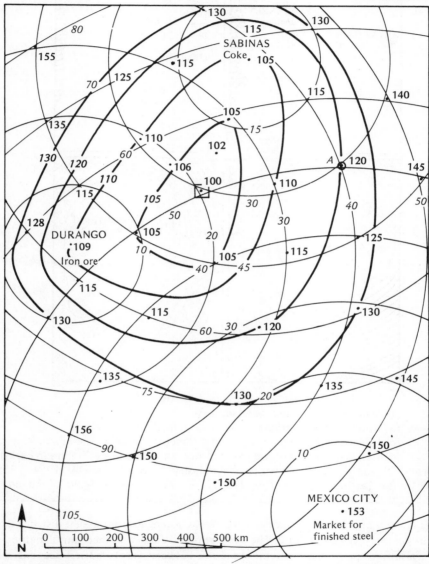

Figure 1.6 Constructing a transport cost surface for the steel industry in Mexico. The light lines are isotims; the heavy lines are isodapanes. The numbers in italics are the values of isotims; the bold numbers are total transport costs (worked out from the isotims); and the bold italic numbers are the values of the isodapanes. All numbers are in pesos. The site with cheapest transport costs is marked by a □.

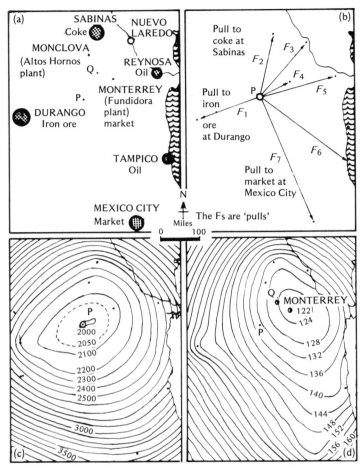

Figure 1.7   The Mexican steel industry in the mid-1950s
   (a) Major material sources and markets
   (b) Application of Varignon's mechanical method
   (c) Isodapanes based on weight and distance (values in tonne-kilometres)
   (d) Isodapanes based on weight, distance, and freight rates (values in pesos)

Adapted from 'The location of the Mexican steel industry' by R. A. Kennelly (1954–1955), *Revista Geografica*, 15, 109–129; 16, 199–213; 17, 60–77: map 4 and figures 7, 10, and 11.

Figure 1.7c is the transport cost surface constructed by Kennelly from weight-distance data. The cheapest site on this surface is point P. The application of Varignon's mechanical method to the weight data also fixed point P as the point of balance between the attractive forces exerted by material sources and by markets (Figure 1.7b). The cheapest site shifted to Paredon, point Q on Figure 1.7a, when constraints imposed by the location of actual transport routes were taken into account. Figure 1.7d is the transport cost surface constructed by Kennelly from actual freight-rate data. The cheapest site on this surface is Monterrey. It so happens that, of the two steel plants in Mexico, the Monterrey plant does in fact incur the lower costs.

2   An entrepreneur, Mr. Ingot, wishes to set up a blast furnace to produce pig iron within the region mapped in Figure 1.8. To produce 1.0 tonne of pig iron, he will need to procure 3.0 tonnes of coal, 2.0 tonnes of iron ore, and 1.0 tonne of limestone. The cost of transporting all raw materials and distributing pig iron to market is £1 per tonne per kilometre.

a   Draw concentric circles (isotims) around the market using a spacing of 1 kilometre. (Isotims for all the raw materials have been drawn for you.)

b   Mark on the values of all isotims. (Remember that it costs £1 to transport one tonne of each material but that different materials are required in different quantities.)

c   Where isotims intersect, and at other convenient points, calculate and mark on the map the total cost of procurement and distribution.

d   Join points of equal transport costs, using a contour interval of £3, to produce an isodapane map. This map will show how total transport costs vary throughout the region.

e   On the basis of the isodapane map alone, where would you advise Mr. Ingot to locate his blast furnace?

f   Study Figure 1.9, which is a map showing railways and some physical features in the region in which the blast furnace is to be located. Do you wish to change your advice to Mr. Ingot? If so, why? In the light of the new information where would you advise Mr. Ingot to build his blast furnace? Give reasons for your decision.

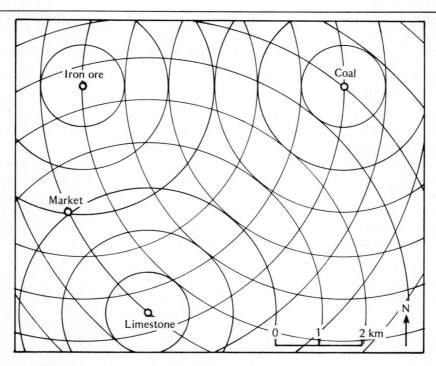

Figure 1.8   Where is the best site for Mr. Ingot's blast furnace?

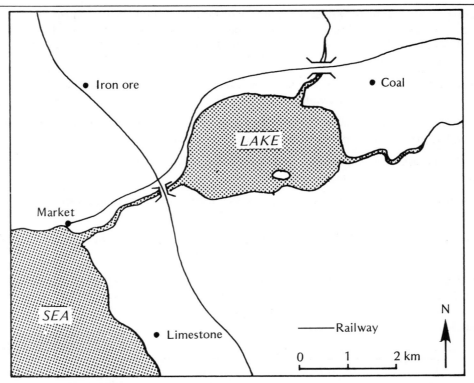

Figure 1.9 Railways and physical features in the region in which Mr. Ingot's blast furnace is to be built

LOCATING A ZINC SMELTER

A zinc smelter produces metallic zinc which is also known as spelter. Metallic zinc is used by metal manufacturers for, among other things, coating iron (galvanizing), in the production of various alloys, the chief of which is brass (a mixture of copper and zinc), and in the making of toothpaste tubes. One of the processes by which metallic zinc is made involves the heating of concentrated ores of zinc in admixture with coal or coke. The zinc ores are reduced to metal which, being volatile, distils and condenses. When mined, the percentage of zinc in the ore is very low, commonly around 3 percent, too low to be fed to a smelter. Zinc ores are therefore concentrated at the site of mining and sent on for smelting. The Broken Hill district of New South Wales, Australia, produces zinc concentrate averaging over 50 percent metallic zinc (Figure 1.10).

Figure 1.11a depicts an imaginary region containing all the raw materials for making metallic zinc. A market for metallic zinc — a brassworks — lies in the southeast of the region. An entrepreneur, Mr. Blende, looks into the possibility of setting up a zinc smelter. The source of zinc ore lies way up north, well away from the sources of coal and from the market. The weights of raw materials required by the smelter per tonne of metallic zinc produced are listed in Table 1.4. The going transport rates for the raw materials and for metallic zinc are also listed in Table 1.4.

The ore would have to be concentrated at the mine because the cost of transporting unconcentrated ore would be very high indeed. The cost of transporting zinc concentrate is high enough. The question

Figure 1.10   A zinc-smelting plant, Broken Hill, Australia. Courtesy of Aerofilms Ltd.

Table 1.4   Weights and rates for the zinc smelter

Material	Weight per tonne of metallic zinc (tonnes)	Transport rate (£ per tonne-kilometre)
Metallic zinc	1.0	0.40
Zinc concentrate	2.0	0.20
Heating coal	2.2	0.22
Reducing coal	0.8	0.22
Fire-clay	0.2	0.10

facing Mr. Blende is where best to set up a smelter. Would it be best in the north by the mine? Would it be best near the brassworks? Or would it be best elsewhere?

Now Mr. Blende is no geographer and he has never heard of Alfred Weber. He gives you a contract to advise him of the best site to build his smelter. What do you do? A first step, and a big one at that, would be to map out the transport costs over the region using the information given in Table 1.4. It would be possible to do this by constructing isotims around each of the four raw material sites and the market and then sketching in the isodapanes. But even for just two raw material sites and a market this procedure is laborious and messy (Figure 1.6 bears witness to this). Instead you use a computer to work out transport costs at a series of grid intersections covering the region (Figure 1.11c) and to produce a contour map of the values for you (Figure 1.11d). But just to check the results, it will pay to follow through the logic of the computations, step by step.

   i   A grid is laid across the map of the region (Figure 1.11b). Steps ii to vii then involve computing transport costs at each grid intersection in turn, taking each to be a potential site for the smelter.

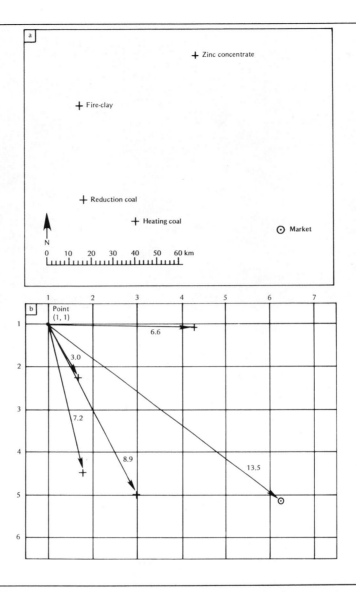

Figure 1.11

ii Start with point (1,1) (Figure 1.11b). The distance from this point to the market is measured as 135 kilometres. (The computer, knowing the coordinates of point (1,1) and of the market, would do this by applying an algebraic formula. You too could use the formula but it is quicker to use a ruler.) Now from a smelter located at point (1,1), every tonne of metallic zinc would have to be transported 135 kilometres to the brassworks at a cost of £0.40 per kilometre. Transporting 1.0 tonne over this distance would cost £54 in all.

iii Next the distance is calculated between point (1,1) and the source of heating coal; it is 89 kilometres. If a smelter were located at point (1,1), for every tonne of metallic zinc produced, 2.2 tonnes of heating coal would have to be procured at a cost of £0.22 per tonne per kilometre. Transporting 2.2 tonnes of heating coal over 89 kilometres would thus cost £43.08.

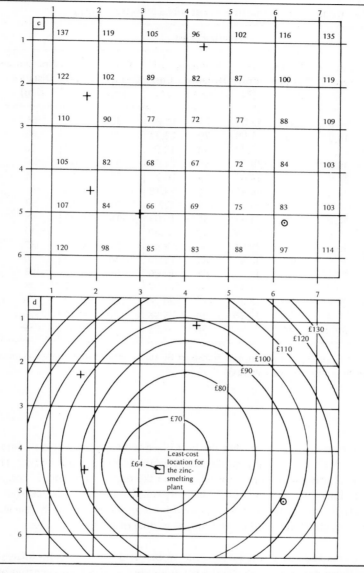

iv The transport costs per tonne of metallic zinc to point (1,1) from all other raw material sources are worked out. The results are listed in row one of Table 1.5.

v All the transport costs to and from point (1,1) are added to give the total transport cost for a smelter located at that point. The sum is £136.75 (Table 1.5, row one).

vi The next point is taken, point (1,2), and steps ii to v repeated using distances appropriate for this point. The results are entered in row two of Table 1.5.

vii The same process is repeated for all grid intersections up to and including point (6,7).

viii The total transport costs for each grid intersection are marked on a map (Figure 1.11d).

ix The values on the map are contoured by joining points of equal total transport costs. (For clarity, the contour map is shown separately here — Figure 1.11d.) These contours are isodapanes and they describe the transport cost surface for zinc-smelting in the region.

Table 1.5 Calculations for constructing a cost surface (extract only)

Point	Transport cost										Total transport costs (£)
	Metallic zinc		Heating coal		Reducing coal		Fire-clay		Zinc concentrate		
	Distance (km)	Cost* (£)	Distance (km)	Cost (£)	Distance (km)	Cost (£)	Distance (km)	Cost (£)	Distance (km)	Cost (£)	
(1,1)	135	54.0	89	43.08	72	12.67	30	0.60	66	26.4	136.75
(1,2)	120	48.0	82	39.69	70	12.32	27	0.54	46	18.4	118.95
(1,3)	106	42.4	80	38.72	74	13.02	37	0.74	26	10.4	105.28
(1,4)	95	38.0	82	39.69	83	14.61	52	1.04	06	2.4	95.74
(1,5)	87	34.8	89	43.08	95	16.72	71	1.42	14	5.6	101.62
(1,6)	83	33.2	100	48.40	109	19.18	90	1.80	34	13.6	116.18
(1,7)	84	33.6	114	55.18	125	22.00	109	2.18	54	21.6	134.56
(2,1)	125	50.0	72	34.85	57	10.03	16	0.32	68	27.2	122.40
(2,2)	107	42.8	63	30.49	50	8.80	09	0.18	49	19.6	101.87
(2,3)	91	36.4	60	29.04	56	9.86	26	0.52	32	12.8	88.62
(2,4)	78	31.2	63	30.49	67	11.79	46	0.92	19	7.6	82.00
(2,5)	68	27.2	72	34.85	81	14.26	66	1.32	23	9.2	86.83
(2,6)	63	25.2	85	41.14	97	17.07	86	1.72	38	15.2	100.33
(2,7)	63	25.6	100	48.40	115	20.24	106	2.12	56	22.4	118.76
(3,1)	116	46.4	56	27.10	34	5.98	20	0.40	76	30.4	110.28
(3,2)	98	39.2	44	21.30	30	5.28	16	0.32	59	23.6	89.70

*This is found by multiplying the distance of transport by the weight of material to be moved (per tonne of product) and the transport rate (see text).

From this information it would seem that the best place to locate Mr. Blende's zinc smelter is at point (4.5, 3.5), at the bottom of the pit described by the transport cost surface. The total transport costs are cheaper here than anywhere else in the region. But would it be wise to advise Mr. Blende to build his smelter at this site? Certainly not. You must first check out the effect other factors, including the pattern of roads, might have on the suitability of the least-cost site for development. You might find that the site is in the middle of a lake!

Assume you have checked out all the factors that Weber held constant in his analysis (p. 9) and you are satisfied that none of them will influence your choice with the exception of the pattern of existing roads and the quality of the land. The existing road network and the quality of the land are mapped in Figure 1.12. The site where transport costs are in theory cheapest lies well off the existing network of roads and in a conservation area. This site cannot therefore be developed. So where is the next-best site? Your computer could be programmed to incorporate the road network as a constraint in the problem and generate a revised transport cost surface. But you decide to test just three sites — A, the bridging point in the west of the region; B, the bridging point just below the market; and C, the site of heating coal.

1 Compute the total transport costs at sites A, B, and C by completing Table 1.6 in conjunction with Figure 1.12. Where would you now advise Mr. Blende to build his zinc smelter? Why?

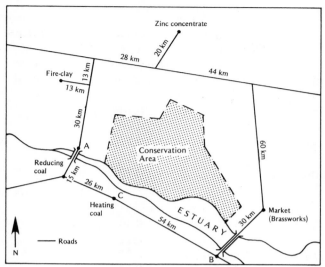

Figure 1.12 Roads and conservation area in the region in which Mr. Blende wishes to build a zinc smelter

THE SWEDISH PAPER INDUSTRY

Transport cost surfaces can reveal some interesting facts about the location of some modern industries. In a painstaking study of the Swedish paper industry, originally published in 1951, Olof Lindberg showed the utility of a cost surface approach. He identified the component costs of procuring spruce pulp-wood, the main raw material, and distributing the paper product. He assumed that road transport rates increased at the same rate in all directions from a paper mill, used actual freight-rate data and existing routes for transport by railway, and also incorporated the effect of waterways used for floating timber. He worked out the cost of procuring spruce pulp-wood for some fifty paper and pulp mills. He also worked out the cost of procuring coal, sulphur, and limestone, other raw materials needed in making paper, and the cost of distributing paper to markets. By adding up distribution and procurement costs, he was able to construct a total transport cost surface for the paper industry in southern Sweden. The surface for the 120,000-cubic-metre level of timber consumption is shown in Figure 1.13. The pattern is complex, showing several sites of relatively low total transport costs. Notice how ports tend to be surrounded by zones of low transport costs. The paper industry is usually regarded as material-orientated: three tonnes of spruce pulp-wood are needed to make one tonne of paper. We should expect to find pulp mills inland, in the heart of the spruce forests where the costs of assembling spruce pulp-wood are cheapest. However, Lindberg found that, because most of the paper is exported, inland mills are at a disadvantage compared with mills at or near the coast (Figure 1.13). In other words, he found that, in this case, the ease and cheapness of links with the export market outweigh the cost of assembling raw materials.

Table 1.6

Site	Material	Weight per tonne of metallic zinc (tonnes)		Transport rate (£ per tonne-kilometre)		Distance of haul* (km)		Cost (£)
	Metallic zinc	1.0	x	0.40	x		=	
	Heating coal	2.2	x	0.22	x		=	
A	Reducing coal	0.8	x	0.22	x		=	
	Fire-clay	0.2	x	0.10	x		=	
	Zinc concentrate	2.0	x	0.20	x		=	_____
						Total cost for site A = £		
	Metallic zinc	1.0	x	0.40	x		=	
	Heating coal	2.2	x	0.22	x		=	
B	Reducing coal	0.8	x	0.22	x		=	
	Fire-clay	0.2	x	0.10	x		=	
	Zinc concentrate	2.0	x	0.20	x		=	_____
						Total cost for site B = £		
	Metallic zinc	1.0	x	0.40	x		=	
	Heating coal	2.2	x	0.22	x		=	
C	Reducing coal	0.8	x	0.22	x		=	
	Fire-clay	0.2	x	0.10	x		=	
	Zinc concentrate	2.0	x	0.20	x		=	_____
						Total cost for site C = £		

*Work this out from Figure 1.12

WEBER'S INDEX OF ORIENTATION

Weber devised an index which could be used to separate those industries where much weight of raw materials (including coal) is lost during processing from those industries where little or no weight of raw materials is lost during processing. This **material index**, as he called it, is defined as

$$\text{Material index} = \frac{\text{Total weight of raw materials (including coal) used per product}}{\text{Weight of the product}}$$

Only those raw materials which tend to occur at particular sites are included in the weight calculations — water, sand, clay, and the like, which are found virtually everywhere, are excluded. Industries which use every bit of raw materials in the production process (these are called **pure materials**) have a material index of 1.0. Industries which use materials that are greatly reduced in weight during the process of production (these are called **weight-loss** or **gross materials**) have a material index in excess of 1.0.

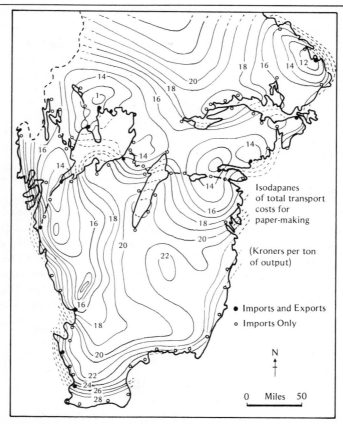

Figure 1.13 Total transport costs for paper manufacturing in southern Sweden. The figures include the cost of transporting spruce pulp-wood, coal, sulphur, limestone, and the paper itself. They are expressed in kroner per ton of output, at a timber consumption of 120,000 cubic metres. The ports are indicated by circles, those open being import harbours and those filled in representing export and import harbours.
Reprinted with permission from 'An economic-geographical study of the localization of the Swedish paper industry' by O. Lindberg (1953), *Geografiska Annaler*, 35, 28—40, figure 20

It may make economic sense to locate a factory in a market when pure materials are used (material index of 1.0 or thereabouts) because the costs of procurement and the costs of distribution will likely be roughly the same. On the other hand, it will almost certainly be cheaper to locate a factory near the source of raw materials when weight-loss materials are used (material index much greater than 1.0) because procurement costs are almost bound to be greater than distribution costs.

1 From the data given in Table 1.7, calculate the material index for each section of the steel industry. For instance, the material index for blast furnaces is 1470/397 = 3.7. What do the values suggest as to the orientation of the different sections of the steel industry?

Table 1.7

Industry	Weight of raw materials (tonnes)	Weight of finished product (tonnes)	Material index
Blast furnaces	1470	397	3.7
Steel mills	119	97	
Tube mills	39	33	
Chain, nail, and screw mills	15	11	
Textile machine shops	5	2.5	

Wilfred Smith tested the efficiency of Weber's material index in 65 British industries. Though the index separated fairly well the industries located at material sources, such as sugar-beet processing in the arable areas of East Anglia, from those which were patently not located at material sources, it by no means provided a flawless method of analysis. The results were improved when the weight of coal was excluded from the calculations (Table 1.8). Even better results were obtained using an index which incorporated the weight of materials per operative. In the engineering industry there is much waste, the weight of materials per operative is surprisingly small, and Weber's index is misleading.

Table 1.8   The material index for some British industries

	Material index (excluding coal)		
	Less than 1.0	1.0 to 2.0	Greater than 2.0
Number of industries located at material source	2	17	3
Number of industries not located at material source	16	14	1

Source: based on W. Smith (1955)

THE EFFECT OF CHEAP LABOUR

Weber realized that in some industries labour is an important factor in production and may exert a strong influence upon location. He suggested that in such cases a site of relatively cheap labour in a region may divert a factory from the site of lowest transport costs. This diversion will take place, argued Weber, if the saving in labour costs exceeds the additional costs of transport incurred in moving to the site of cheap labour supply. In the case of zinc-smelting just described, assume there is a supply of cheap labour at point L_1 (Figure 1.14). At this point, labour is 25

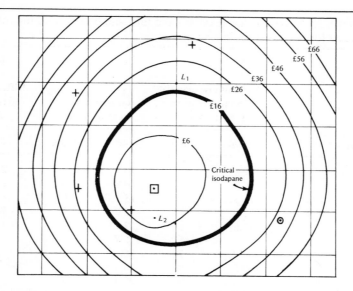

Figure 1.14

percent cheaper than elsewhere, which represents a saving of, we shall say, £16 in total costs per tonne of metallic zinc. In Figure 1.14, the isodapanes join points of equal additional transport cost; they are derived from the isodapanes in Figure 1.11d simply by subtracting the transport costs at the least-cost location, in this case £64, from each isodapane. The £16 additional transport cost isodapane is drawn with a heavy line. Using Weber's words, it is a **critical isodapane**. For a factory located on the critical isodapane, the savings made by using cheap labour, assuming there were a supply of cheap labour somewhere along the critical isodapane, would be just offset by the additional transport costs incurred in moving there. Point L_1, the supply point of cheap labour, lies outside the critical isodapane. If the zinc-smelting plant were to move to the cheap labour supply, the savings it would gain on its labour bill would be eaten up by additional transport costs. Indeed, the savings on labour would be less than the additional transport costs and the plant would run less profitably than it would at the point of lowest transport costs. If, however, the site of cheap labour were at point L_2, which lies within the critical isodapane, the additional transport costs incurred in moving to point L_2 would be less than the savings made in using cheap labour, to the extent of £6 per tonne of metallic zinc. Under these circumstances, the diversion from the point of lowest transport costs to the point of cheap labour supply would be worthwhile.

Weber found that some industries were more susceptible to differences in labour costs from one place to another than others. He computed an index of labour cost which, for any industry, measures the average cost of labour needed to produce a unit weight of product, say a tonne. But of greater importance in assessing the pull of labour, thought Weber, is the ratio of labour cost per tonne of product to the total weight of materials and product to be transported; this ratio is called the labour coefficient and it gives an indication of the susceptibility of an industry to regional variations in the cost of labour.

P. W. Lewis showed that the relative cost of labour increases in the latter stages of production; in other words, labour costs make up a bigger portion of a firm's budget in manufacturing industries, and especially in the final stages of manufacturing, than in primary industries.

Table 1.9 The ratio of the cost of raw materials and fuel, and of wages, to the total value of products for major British industries

1	2	3	4	5	6	7
	Ratio of raw material and fuel costs to total value		Ratio of wages to total value		Difference in rank, d	Difference in rank squared, d^2
Industry	Percent	Rank	Percent	Rank		
Blast furnaces	77.6	1	10.3	22	−21	441
Nonferrous	76.1	2	10.8	20	−18	324
Leather	74.5	3	13.3	16	−13	169
Steel sheets	72.0	4	12.0	19	−15	225
Textiles	71.6	5	12.7	17.5	−12.5	156.25
Chemicals	67.6	6	12.7	17.5		
Iron and steel	64.2	7	16.9	13		
Paper and board	63.5	8	10.5	20		
Clothing	61.8	9	22.4	12		
Stationery	61.0	10	15.4	15		
Cardboard boxes	58.8	11	16.2	14		
Wood and cork	58.6	12	25.0	9		
Precision	58.1	13	23.0	10		
Vehicles	56.7	14	26.8	8		
Electrical engineering	54.7	15	28.9	7		
Other metal	54.6	16	22.6	11		
Food and drink	50.3	17	7.3	23		
Iron foundries	42.9	18	33.3	2		
Shipbuilding	42.4	19	33.6	1		
Treatment of nonmetalliferous	41.9	20	29.4	6		
Engineering	41.3	21	30.4	4		
Printing	31.9	22	31.7	3		
Newsprint	29.3	23	29.8	5		

$\Sigma d^2 =$

The data are used with permission from 'A Numerical Approach to the Location of Industry' by P. W. Lewis (1969), *Occasional Paper Series*, No. 13, University of Hull Publications, table 1, p. 17.

1a Affirm or refute Lewis's findings by constructing a scatter-graph (Figure 1.15) and calculating Spearman's rank correlation coefficient for the data listed in Table 1.9. To calculate the rank correlation coefficient, r, complete Table 1.9 in the following way:

i Work out differences in rank, d, by subtracting values in column 5 from values in column 3. Put the results in column 6.

ii Square the differences in rank and put the results, d^2, in column 7. (N.B. the square of a negative number is a positive number.)

iii Add up the values in column 7 and write the answer at the bottom of the column. This sum is written as Σd^2. The Greek sigma, Σ, means 'sum of' and so the expression Σd^2 means 'the sum of the differences squared'.

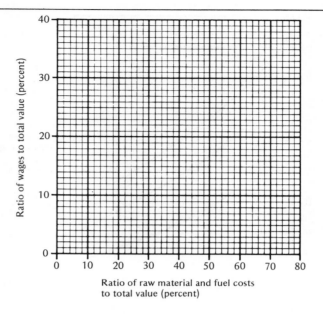

Figure 1.15

Now substitute the value of Σd^2 from Table 1.9 in the formula

$$r = 1 - \left\{ \frac{6 \times \Sigma d^2}{(n^3 - n)} \right\}$$

where n is the number of ranks being compared (the sample size). In the example, n is 23. So we have

$$r = 1 - \left\{ \frac{6 \times \Sigma d^2}{(23^3 - 23)} \right\}$$

$r = $ _____ ?

The value of r will lie between +1.0 and −1.0 (at least it will if your calculations are correct!). A high, positive value indicates a direct relationship between the two sets of ranked data being compared — high ranks in one set correspond to high ranks in the other set, and low ranks in one set correspond to low ranks in the other. A high, negative value indicates an inverse relationship between the two sets of ranked data being compared — high ranks in one set correspond to low ranks in the other set, and vice versa. A value near zero indicates little or no relationship between the sets of ranked data.

Figure 1.16 may be used to test the **statistical significance** of your result. Find the value corresponding to n on the horizontal axis. Now find the value corresponding to your r value on the vertical axis. Locate on the graph the point where these two coordinates intersect. The likelihood of the correlation between the two sets of ranked data arising by chance may now be assessed. If the likelihood is less than 5 percent, you should assume that there is no significant relationship between the two sets of data. If the likelihood is more than 5 percent, you may say that in 95 cases out of 100 the relationship is a result of factors other than chance association.

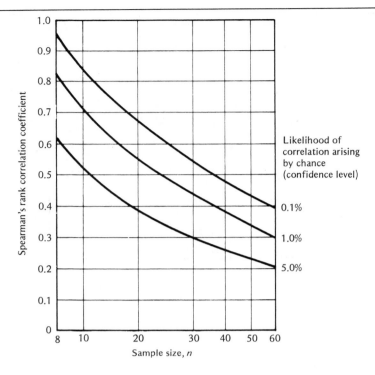

Figure 1.16   Significance values for Spearman's rank correlation coefficient
Based on data in table 6.4 from *Statistics Tables* by H. R. Neave (1978), published by George
Allen & Unwin

b Does your value of *r* suggest a significant relationship between the two sets of data or not? Is the
relationship, if you think there is one, direct or inverse? (Your scatter-graph, Figure 1.15, will
help you to decide this.) Do your findings enable you to affirm or refute Lewis's hypothesis?
Give reasons for your answer.

R. C. Estall and R. O. Buchanan have found that variations in labour costs from one place to
another can be appreciably large. Examining US Bureau of Labour statistics, they discovered
that, in 1970, the average hourly earnings of manufacturing workers in Mississippi were 70
percent lower than workers in Michigan, a difference explained only in part by differences in
industrial structure.

2a Using the data given in Table 1.10, calculate the average hourly labour costs for manufacturing
industry in England, Scotland, and Wales.

b For each standard region, calculate the percentage difference between the national average
labour cost and the regional labour cost.

c Construct a choropleth map (Figure 1.17) to show the percentage difference between national
and regional labour costs.

d Describe and comment on the pattern you have mapped.

Table 1.10   Labour costs for manufacturing industry in England, Scotland, and Wales

Standard region	Total labour costs (£ per hour)	Total labour costs as a percentage of the average national labour costs	The percentage difference between national and regional labour costs
North	1.65	103*	+3†
Yorkshire and Humberside	1.53	96	−4
East Midlands	1.48	93	−7
East Anglia	1.53		
South East	1.73		
South West	1.58		
West Midlands	1.61		
North West	1.59		
Wales	1.73		
Scotland	1.53		
National average	1.60		

*(1.65/1.60) x 100
†103% − 100%

THE EFFECT OF AGGLOMERATION

Weber regarded agglomeration, like labour supply, as a factor which may divert a factory from the site of cheapest transport costs. Imagine five factories producing electronic equipment (Figure 1.18). Each factory occupies the site of cheapest transport costs in its own locational triangle. Assume that a factory can save £5 per tonne of output if it locates near to two others. However, to benefit from this agglomeration, no one factory must incur additional transport costs of more than £5. In Figure 1.18, critical isodapanes of £5 are drawn around each factory. The areas where just two critical isodapanes overlap, shown by light shading, are not favourable for agglomeration because at least three factories must agglomerate for any reduction in production costs to be made. Where three critical isodapanes overlap, as shown by heavy shading, additional transport costs are more than offset by the gains of agglomeration, and it would be worthwhile for factories A, B, and C to move near to one another. But of course, and this is not emphasized in Weber's analysis, it is no use one firm moving to the heavily shaded area on its own — the boards of all three firms must meet and make a decision to move at the same time.

External economies of agglomeration, and links between industries, have become important in determining the location of factories. Savings from external economies of agglomeration can in many industries compensate for higher costs of transport, labour, and other factors of production. In Britain, the growth of industrial estates since 1945 has resulted from external economies of agglomeration. Small firms which in an earlier industrial landscape would have been dispersed, now cluster on industrial estates where they share the costs of public utilities, financial services, and the like and so benefit from reduced operating costs.

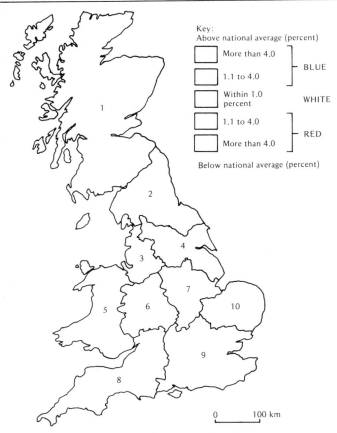

Figure 1.17   Regional variations of labour costs in England, Scotland, and Wales. Regions:
1 Scotland; 2 North; 3 North West; 4 Yorkshire and Humberside; 5 Wales; 6 West Midlands;
7 East Midlands; 8 South West; 9 South East; 10 East Anglia

A classic example of agglomeration is the 'swarming' of the metal industries in the Birmingham
and Black Country conurbation. P. S. Florence has found these metal industries are linked in four
ways. Firstly, there are vertical links, as between the refining of nonferrous metals and the making
of nonferrous hardware goods. Secondly, there are convergent links, as in the production of bolts,
car bodies, tyres, and so forth, all feeding into the motor assembly industry. Thirdly, there are
diagonal links, as seen in toolmakers serving many local industries. And fourthly, there are
indirect links, as in food industries predominated by female workers which balance the dominantly
male workforce of the metal industries.

THE IDEAS OF EDGAR HOOVER AND TORD PALANDER

Weber's analysis was criticized by the Swedish economist Tord Palander in a little-known work,
Contributions to the Theory of Location, published in 1935. Palander's ideas were taken up,
enlarged upon, and given a wider exposure by the American economist Edgar M. Hoover. In his

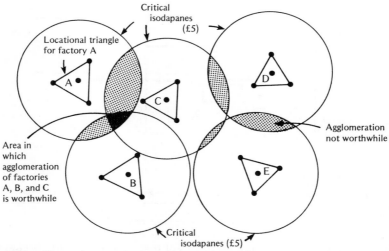

Figure 1.18   Five factories producing electronic equipment

study of the shoe and leather industries, published in 1937, Hoover joined Palander in taking issue with the stress Weber gave to points of least transport cost within a locational triangle. It is unlikely, argued Hoover, that the cheapest site will be anywhere but at one of the material sources or at the market: it is more likely that one of the corners of the triangle will have a pull exceeding the pull of the other two put together. Where a factory is located away from material sources and market the indication is, suggested Hoover, that the location of the industry is not influenced by transport costs at all, and other factors, such as cheap or skilled labour, come into play. In conclusion, Hoover maintained that in practice transport costs are lowest at markets, at sources of raw material, or at breakpoints, such as ports, in the transport network. Hoover developed the logic behind this conclusion in his second book, published in 1948 under the title *The Location of Economic Activity*. We shall look in some detail at Hoover's arguments.

Hoover recognized that transport costs usually increase through a series of steps (Figure 1.19). In many countries, transport rates for a particular mode of transport — rail, road, or water — are divided into a number of zones, the cost of transport tending to decrease with increasing length of haul. In other words, whereas it would cost, say, £20 to transport a load of goods 10 kilometres, it might cost only £30, and not £40, to transport the same load of goods 20 kilometres.

The cost of transport consists of two parts: the cost of loading the goods (**terminal charges**), and the cost of moving the goods (**haulage costs**). Terminal charges and haulage costs are different for rail, road, and water transport (Figure 1.19). Water transport has high terminal charges but low haulage costs and is best suited to long hauls. Road transport has low terminal charges but high haulage costs and is best suited to short hauls. Hoover made a study of freight haulage by 30-ton railway wagons, 10-ton lorries, and package freight barges in the lower Mississippi valley in 1940. He found that road transport was cheapest for hauls of up to 56 kilometres, rail transport was cheapest for hauls between 56 and 608 kilometres, and water transport was cheapest for hauls longer than 608 kilometres.

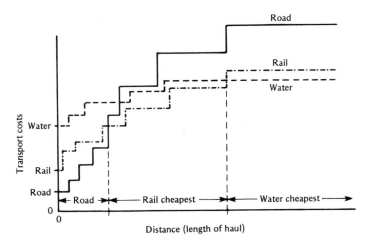

Figure 1.19   Competition between three modes of transport. Road transport has the cheapest terminal charges but the freight rate increases rapidly with increasing length of haul. Water transport has the most expensive terminal charges but the freight rate increases slowly with increasing length of haul. The terminal charges and freight rates for rail transport lie between those for water and road. These differences in the structure of transport costs lead to road being the cheapest form of transport for short hauls, rail the cheapest for medium hauls, and water for long hauls.

Examine a case where a factory uses a single raw material from one source to produce one product sold at a single market. Figure 1.20a shows the variations in procurement and distribution costs for all possible locations between the source of raw material and the market. The best location for the factory is the one involving the smallest total transport costs (procurement plus distribution); in the case pictured it is clearly the source of raw material. In fact, the 'shape' of the transport costs usually dips towards both source of raw material and market (Figure 1.20a and b). It follows from this that the best site for a factory with a single source and a single market will usually be found at either the raw material source or at the market and only exceptionally at an intermediate point.

Where the weight of raw materials is much larger than the weight of product, the factory is likely to locate near the raw material source — it is said to be **material-orientated**. This occurs where combustion or waste of part of the raw material during processing leads to a considerable weight loss; ore-smelting and the crushing of sugar-cane are examples. Also, processes with large fuel requirements — metallurgy, cement-making, glass — involve a high proportion of weight loss and are likely to be found near sources of fuels or other materials. If the relative weights of materials and products are roughly equal but, for some reason, procurement costs per tonne-kilometre are greater than distribution costs, as in food-preserving processes and the ginning and baling of cotton, a material orientation will be the best solution. In general, the early stages of production involve bulk reduction, purifying, or preserving and are material-orientated.

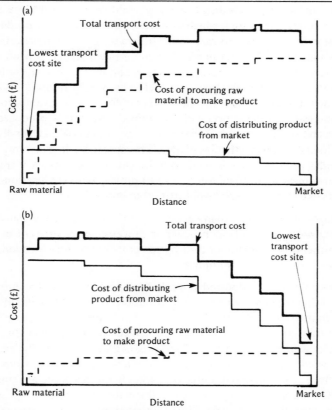

Figures 1.20a and b E. M. Hoover's analysis of (a) material and (b) market locations of industry arising through different combinations of procurement and distribution costs

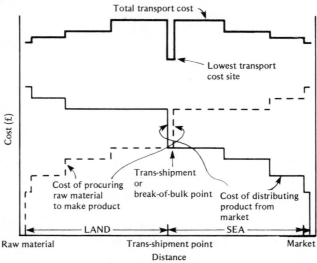

Figure 1.20c E. M. Hoover's analysis of trans-shipment point as a site of lowest transport costs

Location at a source of raw materials then is likely where there is a weight loss in the processing or transfer costs per tonne-kilometre are higher for raw materials than for products. By the same argument, orientation towards markets may result from a weight gain in processing or higher transport costs per tonne-kilometre for products than for raw materials (Figure 1.20b). Raw materials on hand virtually anywhere, such as water, need not enter into the entrepreneur's reckoning of transport costs. They do, however, add to the weight of products like beer and ink and so put up distribution costs; in this case a market location is advantageous. Beverages containing large amounts of water — beer and soft drinks — are usually produced in market sites whereas wines and spirits are made nearer the source of materials. Generally speaking, in the later stages of industrial production, as materials become more fragile, more cumbersome to pack and to handle, more valuable in relation to weight, and progressively graded by size and type, and as customers want small orders at short notice, a market location is advantageous. Perishable products like ice cream also favour **market-orientated** sites.

At intermediate stages of production many goods are at their most readily transferable. Industries concerned with these halfway production processes are indifferent to material or market sources; neither procurement nor distribution costs are decisive locational factors for such industries which are said to be **footloose**.

The industries so far looked at are located at market or material source. Exceptions do arise. Sometimes the variations of procurement and distribution costs lead to a dip in the total transport costs at an intermediate position (Figure 1.20c); this usually crops up where two different modes of transport — water and road, say — meet at a **trans-shipment** or **break-of-bulk point** and the relative weights of raw material and product are similar. In the example sketched in Figure 1.20c, it would be costly to transfer the product or raw material from sea to road (or vice versa) to a factory at raw material source or market because loading and unloading charges can be very high. The cheapest solution is to carry the material by road to a processing plant at the **trans-shipment point**, then send the product to market by sea. This explains why ports and railheads become important manufacturing centres.

Cost Theories in the Modern World

Many important changes have taken place in industry since Weber's day. Improved production techniques have resulted in a reduction in weight loss. For instance, much waste material is usually removed from low-grade metal ores at the site of extraction, a process called **beneficiation**. This means that today a smaller proportion of worthless waste material is transported to a processing plant. Remote sites are now more accessible — changes in transport, particularly in the hauling of bulk cargoes, have rendered economical the long-distance movement of certain materials, crude oil for instance. The advent of electrical power and gas has reduced the dependence of industry on coal. The supertanker, electricity transmission grids, natural gas, and oil have meant that energy-demanding, heavy industries are no longer tied to coalfields as they were during the nineteenth and early twentieth centuries. Modern industries have multifarious requirements. Consider how many components go into a car — Volkswagen claim their 'Beetle' has 998 parts! This has lead to a decline in the location 'pull' of any one material.

The result of all these changes is a tendency for modern industries to favour a market location. Even so, a number of industries do remain tied to sources of raw material. We shall examine some of them.

BRICK-MAKING

Prior to 1945, there were over 1100 brickworks scattered around the United Kingdom using clays of various kinds and serving local markets. By 1973 there were 544 brickworks. The trend has been towards fewer brick-making plants with larger market areas. Today, Fletton bricks, which are made from Lower Oxford Clay, account for 42 percent of the national output. Bricks made from other materials, including silicate, concrete, and other clays, make up the remaining 58 percent.

1a Using the data in Table 1.11, calculate each region's share of national brick production. Work out the mean regional share of national production.

b On Figure 1.21a, shade in black those regions which have an above-average share in the national production of bricks and shade in red those regions which have a below-average share in the national production of bricks.

c Briefly explain the pattern you have mapped.

Another way of assessing the importance of brick-making in different regions is to compute and map regional location quotients. A **location quotient** is defined by this formula:

$$\text{Location quotient} = \frac{\dfrac{\text{Number of employees in industry } i \text{ in region } j}{\text{Number of employees in all industry in region } j}}{\dfrac{\text{Total number of employees in industry } i \text{ in the nation}}{\text{Total number of employees in all industry in the nation}}}$$

Table 1.11   Brick production in England, Scotland, and Wales, 1978

Region	Millions of bricks	Percent share of national production
North	318	
Yorkshire and Humberside	277	
East Midlands	341	
East Anglia	697	
South East	1789	
South West	199	
West Midlands	292	
North West	381	
Wales	142	
Scotland	400	
Totals	4836	

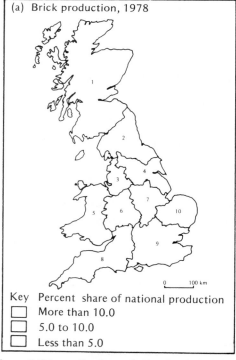

(a) Brick production, 1978

Key   Percent share of national production
- More than 10.0
- 5.0 to 10.0
- Less than 5.0

(b) Brick-making, 1976

Key   Location quotients
- More than 1.5
- 1.0 to 1.5
- Less than 1.0

Figure 1.21   Regions: 1 Scotland; 2 North; 3 North West; 4 Yorkshire and Humberside;
5 Wales; 6 West Midlands; 7 East Midlands; 8 South West; 9 South East; 10 East Anglia

A location quotient of 1.0 means that the industry in question, in terms of the workforce it employs, is represented in the region in question in the same proportion as the national average. A value greater than 1.0 indicates that the industry is concentrated in the region relative to the nation; a value less than 1.0 indicates that the industry is underrepresented in the region relative to the nation.

2a By completing Table 1.12, work out location quotients for brick-making in the regions of the United Kingdom.

b Construct a choropleth map of the location quotients (Figure 1.21b).

c Describe and discuss the pattern you have mapped.

Table 1.12 Data for calculating regional location quotients for brick-making

1	2	3	4	5
	Number of people employed in all manufacturing industries (thousands)	Number of people employed in brick-making (thousands)	$\dfrac{\text{Column 3}}{\text{Column 2}}$	Location quotient: Column 4 $\left(\dfrac{\text{Total of column 3}}{\text{Total of column 2}}\right)$
Region				
South East	1851.0	7.5		
East Anglia	195.8	No data	—	—
South West	419.9	No data	—	—
West Midlands	978.7	No data	—	—
East Midlands	587.1	5.5		
Yorkshire and Humberside	711.3	6.5		
North West	1005.7	3.1		
North	438.1	2.7		
Wales	302.7	1.3		
Scotland	607.8	4.4		
Totals*	7098.6	39.8		

*The totals are for Great Britain and may exceed the sum of the figures listed.
Source of data: Annual Census of Employment, 1976, *Department of Employment Gazette*, December 1977

3 Like many raw material-orientated industries, brick-making is sensitive to transport costs which account for 16 percent or so of total costs. Brick plants endeavour to keep transport costs as low as possible, as the data in Table 1.13 would suggest. Short-haul trips being the cheapest, it

Table 1.13 Brick deliveries by road

Road mileage from plant	Percentage of total deliveries
Less than 50	48
50 to 100	39
More than 100	13

would seem to be of advantage for a brickworks to be located near a market. Clarify this apparent contradiction between the tendencies to both market orientation and material orientation shown by brickworks. (Bear in mind the effect internal economies of scale might have — p. 7.)

CEMENT-MAKING

The cement industry shares a number of characteristics with the brick-making industry: its product is of low unit value and is bulky; its raw materials (limestone and clay or mud) are quarried; its market is the construction industry; and its transport costs are a relatively big portion of total costs — 12 percent.

1 a   Complete Table 1.14 and then, using the same procedure as for brick-making, make a choropleth map (Figure 1.22) of regional location quotients for cement-making.

b   Compare and contrast the regional patterns of brick-making and cement-making as revealed by the location quotients.

2   The material index for cement-making is high. To what extent is the high material index reflected in the distribution of cement works as mapped in Figure 1.23a? (To answer this question you will need to identify the source of limestone for each cement works.)

3   The Pitstone cementworks is located near the Tring Gap at the foot of the Chiltern Hills (cf. Figure 1.24). Why is the Tring Gap a good place to make cement? (Consider the role of transport, raw materials, and the markets as indicated on Figure 1.23b.)

Table 1.14   Data for calculating regional location quotients for cement-making

1	2	3	4	5
Region	Number of people employed in all manufacturing industries (thousands)	Number of people employed in cement-making (thousands)	$\dfrac{\text{Column 3}}{\text{Column 4}}$	Location quotient: $\dfrac{\text{Column 4}}{\left(\dfrac{\text{Total of column 3}}{\text{Total of column 2}}\right)}$
South East	1851.0	10.0	0.0054	0.59
East Anglia	195.8	1.0	0.0051	0.56
South West	419.9	0.8	0.0019	0.21
West Midlands	978.7	6.4	0.0065	0.71
East Midlands	587.1	3.4	0.0058	0.64
Yorkshire and Humberside	711.3	12.4	0.0174	1.91
North West	1005.7	19.4	0.0196	2.15
North	438.1	5.9		
Wales	302.7	1.8		
Scotland	607.8	3.7		
Totals*	7098.6	64.9	0.0091	

*Totals are for Great Britain and may exceed the sum of the figures listed.
Source of data: Annual Census of Employment, 1976, *Department of Employment Gazette*, December 1977

Figure 1.22 Regions: 1 Scotland; 2 North; 3 North West; 4 Yorkshire and Humberside; 5 Wales; 6 West Midlands; 7 East Midlands; 8 South West; 9 South East; 10 East Anglia

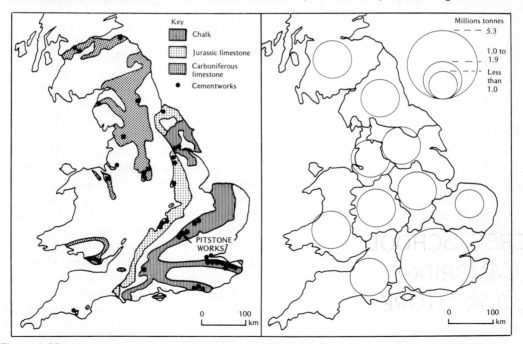

Figure 1.23a Cement works

Figure 1.23b Cement deliveries in 1971

4   A number of cement manufacturers have set up strategically located depots throughout the country which are served by specially designed railway cement wagons. Does this fact tie up with Hoover's observation that railways are the cheapest mode of transport over medium distances (p. 33)? Give reasons for your answer.

Figure 1.24   Chinnor cementworks at the foot of the Chiltern chalk scarp. Courtesy of Aerofilms Ltd.

IRON- AND STEEL-MAKING

Historical Background

The iron and steel industry in Britain has over the centuries shifted its location, changed its modes of production, and depended on ore supplies from different sources at different times. The result is today a complicated pattern of distribution (Figure 1.25). Originally, iron was smelted in charcoal furnaces in areas where iron ore and wood were both at hand. At a later stage, water power was harnessed to drive bellows and hammers at, or near, these original sites in places like the Weald of Kent and the Forest of Dean. In the eighteenth century coke was used for smelting. This removed the need to locate near supplies of timber and in areas where water power was available and led to a shift from forest sites to coalfields. Some coalfields, like the one in West Yorkshire, were notably attractive because iron ore was interbedded with coal and thus the costly business of procuring iron ore or coal could be avoided. Large-scale iron-producing regions did not develop until 1879 when the invention of the Gilchrist-Thomas process enabled home supplies of low-grade phosphoric ore to be exploited. Before 1879, when home ore was not used on a large scale, the sites where foreign ore was imported were attractive locations for iron smelters, especially as the overseas market for iron developed. More recent technological

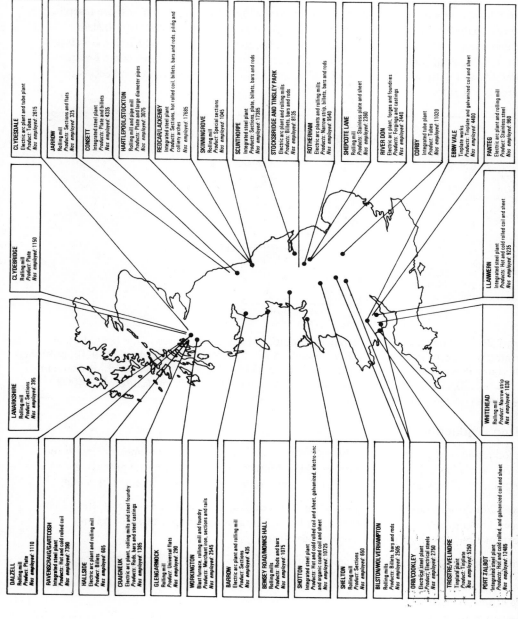

DALZELL
Rolling mill
Product: Plate
Nos employed: 1110

RAVENSCRAIG/GARTCOSH
Integrated steel plant
Products: Hot and cold rolled coil
Nos employed: 7380

HALLSIDE
Electric arc plant and rolling mill
Product: Billets
Nos employed: 605

CRAIGNEUK
Electric arc plant, rolling mills and steel foundry
Products: Rods, bars and steel castings
Nos employed: 1385

GLENGARNOCK
Rolling mill
Product: Universal flats
Nos employed: 290

WORKINGTON
Blast furnace, rolling mill and foundry
Products: Merchant iron, sections and rails
Nos employed: 2545

BARROW
Electric arc plant and rolling mill
Product: Sections
Nos employed: 435

BENSEY ROAD/MONKS HALL
Rolling mills
Products: Rods and bars
Nos employed: 1075

SHOTTON
Integrated steel plant
Products: Hot and cold rolled coil and sheet, galvanized, electro-zinc and organic coated coil and sheet
Nos employed: 10725

SHELTON
Rolling mill
Product: Sections
Nos employed: 650

BILSTON/WOLVERHAMPTON
Rolling mills
Products: Billets, bars and rods
Nos employed: 2505

ORB/COOKLEY
Electrical steel plant
Product: Electrical steels
Nos employed: 2750

TROSTRE/VELINDRE
Tinplate plant
Product: Tinplate
Nos employed: 5250

PORT TALBOT
Integrated steel plant
Products: Hot and cold rolled, and galvanized coil and sheet
Nos employed: 12485

WHITEHEAD
Rolling mill
Product: Narrow strip
Nos employed: 1030

LLANWERN
Integrated steel plant
Products: Hot and cold rolled coil and sheet
Nos employed: 9235

CLYDEBRIDGE
Rolling mill
Product: Plate
Nos employed: 1150

LANARKSHIRE
Rolling mill
Product: Sections
Nos employed: 395

CLYDESDALE
Electric arc plant and tube plant
Product: Tubes
Nos employed: 2615

JARROW
Rolling mill
Products: Sections and flats
Nos employed: 325

CONSETT
Integrated steel plant
Products: Plate and billets
Nos employed: 4335

HARTLEPOOL/STOCKTON
Rolling mill and pipe mill
Products: Plate and large diameter pipes
Nos employed: 3075

REDCAR/LACKENBY
Integrated steel plant
Products: Sections, hot rolled coil, billets, bars and rods, piling and colliery arches
Nos employed: 17685

SKINNINGROVE
Rolling mill
Product: Special sections
Nos employed: 1045

SCUNTHORPE
Integrated steel plant
Products: Sections, plate, billets, bars and rods
Nos employed: 17385

STOCKSBRIDGE AND TINSLEY PARK
Electric arc plant and rolling mills
Products: Billets, bars and rods
Nos employed: 8135

ROTHERHAM
Electric arc plants and rolling mills
Products: Narrow strip, billets, bars and rods
Nos employed: 9540

SHEPCOTE LANE
Rolling mill
Products: Stainless plate and sheet
Nos employed: 2360

RIVER DON
Electric arc plant, forges and foundries
Products: Forgings and castings
Nos employed: 2440

CORBY
Integrated tube plant
Product: Tubes
Nos employed: 11020

EBBW VALE
Tinplate works
Products: Tinplate and galvanized coil and sheet
Nos employed: 4400

PANTEG
Electric arc plant and rolling mill
Product: Stainless steel
Nos employed: 960

Figure 1.25 British Steel Corporation: main iron and steel works, products, and numbers employed (31 March 1979)
Reprinted with permission from *The British Steel Corporation, Annual Report and Accounts 1978–79*

44

developments have reduced the amount of coal needed to produce a tonne of pig iron to one-third of its value in the nineteenth century. This has helped to reduce the very strong 'pull' which coalfields exerted on many industries.

A similar sequence of events has taken place in the United States. Originally, the forested area of eastern Pennsylvania became an important iron-producing region. When the iron ore reserves in Minnesota could be exploited, lake ports such as Chicago, Cleveland, and Detroit became profitable sites; in addition to being near to coalfields, they were major industrial markets and lay at trans-shipment points. Since 1940, expansion has occurred in the large coastal markets of Philadelphia and Baltimore which have large quantities of scrap for recycling and access to imported ore (Figure 1.26).

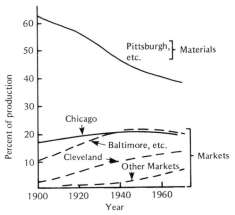

Figure 1.26   The changing importance of some iron and steel centres in the United States Reprinted with permission from *Spatial Organization of Society* by R. L. Morrill (1970), published by Duxbury Press, North Scituate, Massachusetts, a division of Wadsworth Publishing Company, Inc., figure 5.20,© 1970 Wadsworth Publishing Company, Inc., Belmont, California, 94002

Pig Iron Production

The production of pig iron involves the combustion of iron ore, coke, and limestone, with some scrap iron and a few minor ingredients, in a blast furnace. Weight is lost in the process, giving a material index of between 3 and 4 (p. 24). The pig iron produced is 95 percent pure metal.

1a   Figure 1.27a shows sources of coking coal, areas producing low-grade iron ore, and sites of blast furnaces. Using Figure 1.27b, depict the number of furnaces at each site (Table 1.15) by proportional circles using the scale given on the figure.

b   Comment on the pattern you have mapped by

   i   assessing the impact of sources of low-grade, home-produced ore and high-grade, imported ore (Table 1.16); and
   ii   examining the extent to which the pattern can be explained by classical theories of industrial location. (Consider the possible effects of sources of raw materials, break-of-bulk points, and the market provided by steel producers as listed in Table 1.17.)

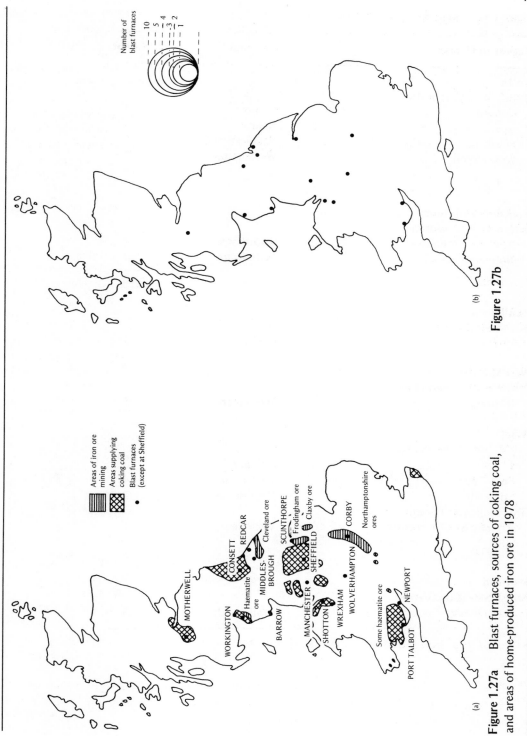

Figure 1.27a   Blast furnaces, sources of coking coal, and areas of home-produced iron ore in 1978

Figure 1.27b

Table 1.15   Blast furnaces, 1978

Region and works	Location	Number
North		
British Steel Corporation		
Consett . . .	Consett	3
Redcar . . .	Redcar	—
South Teesside — Cleveland .	Middlesbrough	5
Workington . . .	Workington	3
Total . . .		11
Yorkshire and Humberside		
British Steel Corporation		
Appleby Frodingham . .	Scunthorpe	7
Normanby Park . . .	Scunthorpe	3
Total . . .		10
East Midlands		
British Steel Corporation		
Corby . . .	Corby	4
West Midlands		
British Steel Corporation		
Bilston . . .	Wolverhampton	1
Wales		
British Steel Corporation		
Llanwern . . .	Newport	3
Port Talbot . . .	Port Talbot	4
Shotton . . .	Deeside	2
GKN Rolled and Bright Steel Ltd.		
Brymbo Steel Works Ltd. . .	Wrexham	1
Total . . .		10
Scotland		
British Steel Corporation		
Ravenscraig . . .	Motherwell	3

Source of data: Iron and Steel Statistics Bureau, 1979

Table 1.16   Iron ore production and consumption, 1978 to 1979

Hematite	Percent Iron	Production (*thousand tonnes*)
Cumberland and Glamorgan	*45*	107.4
Jurassic		
Lower lias — N. Lincs	*21*	1,837.9
Inferior oolite — S. Lincs, Leics, Northants, and Rutland	*29*	2,105.8

Consumption in blast furnaces, sinter plants, and steelworks

Iron ore	
Home	4,051.1
Imported	15,768.2

Source of data: Annual Statistics for the Corporation, 1978–1979, British Steel Corporation

Steel-Making

Steel production involves the combustion of pig iron and scrap (with alloys of iron thrown in), iron ore, and limestone. Three methods of production are currently in use: the oxygen converter, open hearth (oil fired), and electric arc (used mainly in the production of high-grade, alloy steel). Both the open hearth and the electric arc methods are designed to use large quantities of scrap metal, unlike the converter method which can use at most a 30 percent charge of scrap. There has, however, been a move away from open hearth furnaces to oxygen converters because oxygen converters can produce steel more cheaply. Costs may also be reduced by integration, by producing iron and steel at the same site (Figure 1.28).

2   Figure 1.29 shows nine 'model' locations for iron and steel plants given the source of ore or scrap, the source of coking coal, and the market. Each of these 'model' locations represents the 'best' site for the iron and steel plant under particular conditions of raw material sources and markets. Table 1.17 shows the location and type of privately owned and state-owned steel furnaces in the United Kingdom. For the sites of steel furnaces listed in Table 1.18, and by answering 'yes' or 'no' to the questions posed about source of iron ore, source of coking coal, and the market, decide which of the nine 'model' locations seems the closest match with the actual location. (Details of sources of iron ore and coking coal are given in broad terms in Figure 1.27a. Bear in mind that home-produced ores are of low grade whereas imported ores are usually high grade. Finished products of steelworks are used as raw materials for many other industries, especially for tinplate manufacture, which is concentrated in the Swansea-Llanelly area; shipbuilding and marine engineering, which are found at Clydeside, Tyneside, Barrow-in-Furness, and Belfast; heavy engineering, found mainly in Middlesbrough, West Yorkshire, and central Scotland; the manufacture of heavy electrical equipment, which is concentrated in the Midlands (Rugby and Coventry in particular), southern Lancashire, and

Table 1.17   Steel furnaces in the United Kingdom in 1979

	British Steel Corporation		Privately owned[†]	Total
	Oxygen converters	Others*		
Barrow	—	1	—	1
Consett	2	—	—	2
Middlesbrough	3	3	—	6
Sheffield	—	11	49	60
Rotherham	—	7	—	7
Scunthorpe	5	—	—	5
Bradford	—	—	6	6
Corby	3	2	—	5
Sheerness	—	—	2	2
Wolverhampton	—	7	—	7
Dudley	—	—	4	4
Wednesbury	—	—	5	5
Birmingham	—	—	1	1
Manchester	—	—	3	3
Birkenhead	—	—	1	1
Newport	3	—	4	7
Pontypool	—	1	—	1
Port Talbot	2	—	—	2
Deeside	—	12	—	12
Llanelli	—	—	1	1
Wrexham	—	—	4	4
Cardiff	—	—	2	2
Motherwell	3	6	—	9
Glasgow	—	1	—	1

*Basic open hearth and electric arc and induction.
†All privately owned furnaces are electric arc or induction furnaces except two
  in Wednesbury which are basic open hearth furnaces.
Source of data: Iron and Steel Statistic Bureau, 1979.

the London area; for the automobile assembly industry; and for textile machinery manufacture
in Manchester and southern Lancashire.)

3a   By completing Table 1.19, work out regional location quotients for the iron and steel industry
      in the United Kingdom.

b   Draw a choropleth map of the location quotients (Figure 1.30).

c   Describe and explain the pattern you have mapped.

d   Since 1979 a number of steelworks have been closed. Find out which they are and suggest reasons
      for the geographical pattern of closures?

Figure 1.28   The Margam Steelworks, Port Talbot. Courtesy of Aerofilms Ltd.

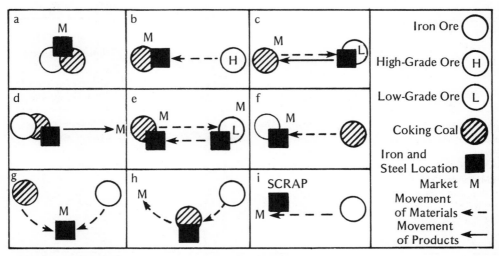

Figure 1.29   Model locations of iron and steel industries
Reprinted with permission from 'Models of industrial location' by F. E. Ian Hamilton (1967) in *Models in Geography*, edited by R. J. Chorley and P. Haggett, published by Methuen, London, figure 10.17

Table 1.18

Site of steel furnaces	Is the site on a coalfield?	Is the site on an iron ore area?	Is the site in a big market?*	'Model' location (Figure 1.29)
Barrow				
Corby				
Deeside				
Manchester				
Middlesbrough				
Motherwell				
Port Talbot				
Wolverhampton				

*If yes, assume scrap is available in large quantities. Does the availability of scrap tie up with the type of steel furnace? Remember, open hearth and electric arc furnaces are designed to take a large charge of scrap metal whereas oxygen converter furnaces are not.

Table 1.19   Data for calculating regional location quotients for iron- and steel-making

1	2	3	4	5
Region	Number of people employed in all manufacturing industries (thousands)	Number of people employed in iron- and steel-making (thousands)	$\dfrac{\text{Column 3}}{\text{Column 2}}$	Location quotient: $\dfrac{\text{Column 4}}{\left(\dfrac{\text{Total of column 3}}{\text{Total of column 2}}\right)}$
South East	1851.0	5.4		
East Anglia	195.8	No data	—	—
South West	419.9	No data	—	—
West Midlands	978.7	24.1		
East Midlands	587.1	6.6		
Yorkshire and Humberside	711.3	70.2		
North West	1005.7	6.4		
North	438.1	36.3		
Wales	302.7	61.6		
Scotland	607.8	20.3		
Totals*	7098.6	232.6		

*The totals are for Great Britain and may exceed the sum of the figures listed.
Source of data: Annual Census of Employment, 1976, *Department of Employment Gazette*, December 1977

Figure 1.30   Iron and steel: regional location quotients in 1976. Regions: 1 Scotland; 2 Northern Ireland; 3 North West; 4 Yorkshire and Humberside; 5 Wales; 6 West Midlands; 7 East Midlands; 8 South West; 9 South East; 10 East Anglia

The Role of Revenue

The influence of markets, which strongly influence revenue, on deciding where to locate an industry has this century become increasingly important. As was noted earlier (p. 37), this is partly due to a general, relative lowering of transport rates, a number of technological developments which have enabled less material to be procured to make a unit weight of product, and a bigger relative drop in the cost of transporting bulky raw materials as compared with the cost of distributing finished products. A trend towards market orientation of industry was observed in the United States during the 1920s. This was associated with the growth of the so-called consumer society.

There are many kinds of market. Households are the **final market** and this market attracts what are known as **residentiary industries**. The size of the labour force in a residentiary industry is commonly directly proportional to the population of the market. **Industrial markets**, in which the production of one industry is the raw material for another, form another category. The car assembly industry is a good example of this (p. 32). Some raw materials create what are called **resource markets**. The classic example of a resource market is the tin-can industry located in fruit- and vegetable-canning districts.

MARKET AREAS AND COMPETITION

One of the first attempts to predict the market area of an industrial enterprise was made by the Swedish economist Tord Palander in his book entitled *Contributions to the Theory of Location*. One of the questions Palander asked was this: Given the site of production (the factory), competitive conditions, factory costs, and transport rates, how does the price of the product influence the area in which an entrepreneur can sell his or her goods, that is, the **market area**? This is Weber's problem in reverse. To answer the question, Palander looked at two firms making the same product. We shall illustrate his argument with two competing breweries.

Two Competing Breweries

In Palander's analysis, the price of beer on delivery would consist of two parts: the first is the price of beer at the brewery and the second part is a delivery or transport charge. The drayer's delivery charges are assumed to increase at a fixed rate with increasing distance from the breweries. The price of beer delivered at pubs from Drown's brewery is shown in Figure 1.31a. At Sorrow's brewery, production costs are a little less and so, therefore, is the price of beer at the plant. The drayer's rates are the same though, giving the pattern of delivered prices shown in Figure 1.31b. Putting Figures 1.31a and b together (Figure 1.31c), we find that the lines representing the delivered price of beer from the two breweries cross. The position at which the lines cross marks the boundary of Drown's and Sorrow's market areas. To the left of the boundary, Sorrow's brewery cannot deliver beer at a lower price than Drown's; to the right of the boundary Sorrow's brewery is unable to compete with Drown's. Notice that because Sorrow's brewery has the lower production costs, its market area is the larger.

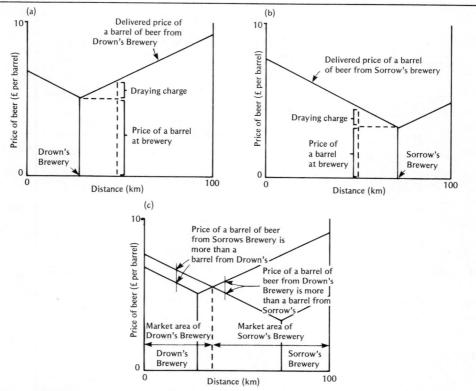

Figure 1.31 Palander's analysis of the market areas of two competing firms illustrated for two breweries

1   Five breweries producing bitter are located at the points labelled A to E in Figure 1.32. The production costs of a pint of bitter at each brewery are 22p at brewery A, 24p at brewery B, 26p at brewery C, 28p at brewery D, and 30p at brewery E. The delivery charges from all breweries are 5p per pint per ten kilometres.

a   Plot lines showing the delivered price of a pint of beer from each of the five breweries. Do this on Figure 1.32.

b   Mark the market areas of each brewery, assuming that the purchasers of beer are evenly spread and will buy the cheapest beer available to them.

c   What would happen to the market areas if brewery D were to cut production costs by 10p per pint?

d   Comment on the relationship between production costs and market area. How would different transport charges alter this relationship?

Competing Ice-cream Sellers

The problem of market areas resulting from competition between two businesses, where the customers are spread along a line such as a road, had also been tackled by F. Fetter in 1924 and Harold Hotelling in 1929. Hotelling considered the case of two ice-cream sellers

54

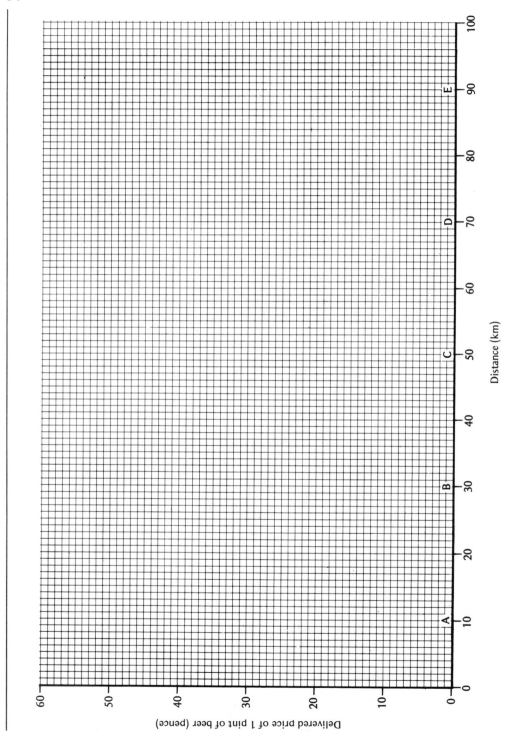

Figure 1.32

supplying an identical product, say ice-cream cornets, to customers spread along a seaside beach. Hotelling's analysis was based on the following assumptions.

i The customers are spread evenly along the beach.

ii The cost of an ice-cream cornet consists of two parts (Figure 1.33a). The first part is the actual price paid for the cornet at the seller's stall. The second part of cost is the effort made by a customer in walking along the beach to buy a cornet. This effort can be translated into 'cost'. Hotelling assumed that the effort of walking along the beach increases with increasing distance from an ice-cream seller. In other words, people sitting near the seller make less effort, and therefore incur less 'cost', than those sitting farther away. (If, instead of two ice-cream sellers, we were considering two competing supermarkets located on a long road, then the effort on the part of the customers in travelling to a supermarket could be measured in terms of real money as bus fares. Clearly, customers living nearer the supermarket would incur smaller bus fares than customers living farther away and, in effect, they would pay less for their goods.)

iii Each customer buys one ice-cream cornet in a unit of time regardless of price (this means that the demand is inelastic — see *Agriculture*, p. 63), but will always buy from the seller nearer to him (or her) because the effective cost of the nearer seller's ice-cream cornet will be lower than that of the competitor (Figure 1.33a).

With these assumptions, Hotelling concluded that the two sellers would end up back-to-back at the centre of the beach, each serving one-half of the market (Figure 1.33a). The argument he used runs like this. An ice-cream seller, let us call him Mr Knickerbocker, comes onto the beach and starts selling at the central point. (In fact, any position would give him the entire market.) A second ice-cream seller, Mr. Glory, comes along. He can start selling wherever he wishes to but he must be prepared to compete with Mr. Knickerbocker. The best thing Mr. Glory can do is to join Mr. Knickerbocker at the centre of the beach and then they will each have a monopoly over one-half of the market. Had Mr. Glory chosen to sell at point B (Figure 1.33b) he would have served the right-hand end of the market at a cheaper price than he would have from the central point. But as demand is assumed to be inelastic, the customers will buy from him whatever the price (so long as he is undercutting Mr. Knickerbocker's price) and he would have gained no particular advantage from doing so. Indeed, if he were to locate at point B, Mr. Knickerbocker could compete with him in the portion of the beach lying between them thus depriving Mr. Glory of some of the market he would have had if he were to have sold from the centre. Standing back-to-back with Mr. Knickerbocker is the only way that Mr. Glory can control half the market. Hotelling argued that a third, and any other, ice-cream seller would also join the central cluster; he extended this finding to the development of industrial agglomeration under certain conditions of demand.

Hotelling's argument has been shown to contain flaws. The two ice-cream sellers do not have to share a central location to command half of the market; they may both set up an equal distance either side of the centre to achieve the same result (Figure 1.33c). If they were centrally placed, the arrival of a third ice-cream seller would lead to jostling for the outside positions because the one in the middle would make few, if any, sales. What is more likely to happen as more and more ice-cream sellers come onto the beach is that they will tend to spread in pairs along the beach. The 'final' solution to the problem was proposed by Michael Bradford and Ashley Kent in their book *Human Geography: Theories and their Applications*; this, the Jaws solution, is shown in Figure 1.33d.

56

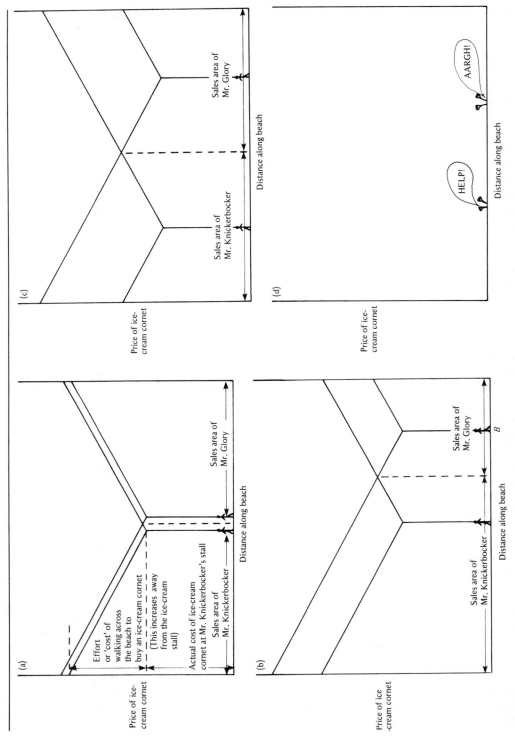

Figure 1.33 The market areas of two ice-cream cornet sellers who compete for business along a beach

2 Two sweet shops are to be located along a street. Assume that the number of shoppers is the same at all points along the street, that the shoppers will always buy from the supplier who offers the cheapest sweets, and that the sweet shops may be located anywhere along the street. Where should the sweet shops locate to capture as much of the market as possible?

MAPPING MARKET AREAS

The Ideas of Edgar Hoover

Palander's pioneering work (p. 52) has never been translated from the German, but many of the seminal ideas in it were taken up and developed by Edgar Hoover in his classic study of the shoe and leather industry which was published in 1937. In like manner to Weber, Hoover showed how the cheapest location in terms of transport costs can be found by constructing isotims around material sources and markets and then drawing lines of equal total transport costs (isodapanes). However, Hoover took the analysis a step further by showing how different parts of the market will be served by different points of production (factories).

Three competing plant locations for electrical goods manufacture in the Great Lakes, northern Georgia, and northern Mississippi are shown in Figure 1.34. The operating costs, that is, the cost of procuring materials to make a hundredweight of output plus the cost of production, are $13.11 in the Great Lakes, $13.82 in northern Georgia, and $13.61 in northern Mississippi.

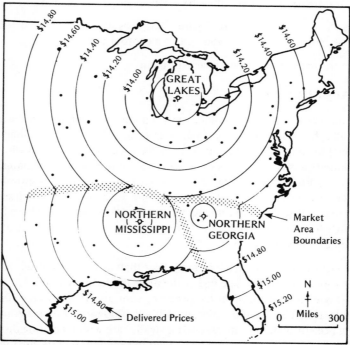

Figure 1.34   The delivered price of electrical goods from three competing plants
Reprinted with permission from *Industrial Location* by D. M. Smith (1971), published by John Wiley & Sons, New York, figure 18.9

The costs of loading a hundredweight of production is $0.50 at all sites. And delivery charges are the same from each site, being $0.0015 per mile. The **delivered price** of electrical goods from any one of the sites can be calculated for all customers by adding delivery costs to operating costs. So a customer located 300 miles from the Great Lakes site would be charged

$13.11   + $0.50   +   ($0.0015   x   300)   = $14.06

Operating   Loading   Transport   Distance
cost   charge   rate

Similarly, a customer located 600 miles from northern Mississippi would be charged

$13.61   + $0.50   +   ($ 0.0015   x   600)   = $15.01

As delivery charges increase in a regular way (0.15 cents every mile), so will the delivered price increase at a constant rate from the production points. At all points 300 miles from the Great Lakes site the delivered price will be the same and can be represented by an isotim. Figure 1.34 shows isotims constructed from all three production sites. The boundaries of the market areas of the three sites are found where isotims of the same value intersect. Study the map and you will see that anywhere north of the boundary between the Great Lakes market area and the northern Mississippi market area it is cheaper to buy from the Great Lakes site, whereas anywhere south of that boundary it is cheaper to buy from the northern Mississippi site.

1   Figure 1.35 shows the location of three competing producers of cement. The operating costs, loading charges, and transport charges of each cement plant are listed in Table 1.20.

a   For each cement plant, plot on the map (Figure 1.35) lines showing points of equal delivered price of a tonne of cement (these are isotims). Use a contour interval of £2.

b   Mark on the map the market area of each cement plant.

The Ideas of August Lösch

Lösch saw demand (market size) as the main control upon the location of industrial activities. To develop this idea, he considered the growth of an economic activity (we shall look at beer-making by farmers) in an imaginary region. The region is a uniform plain on which raw materials are evenly distributed (this is the reverse of Weber's assumption about raw materials) and on which transport rates are everywhere the same. The farming population is spread evenly, all farmers having identical tastes, technical know-how, and economic opportunities, and living in evenly distributed, self-sufficient farms. At the outset, all farmers make beer for their own consumption. But what will happen if some farmers produce a surplus of beer to sell? The answer is, according to Lösch, like this:

Assume just one farmer decides to produce more beer than he can use himself and so has a surplus which he can sell to others. The higher the price the farmer charges for his beer, the less of it he will sell (Figure 1.36a), all other factors being constant. If he charges 30p a pint, he will sell 2000 pints. If he charges 40p a pint, he will sell 1000 pints. Lösch assumed that the cost of delivering beer to customers increases at a steady rate with increasing distance from the farm; and that the delivery costs would be added to production costs and met by the customers, not by the farmer. This means the price of beer increases away from the farm. The higher the price the fewer the sales — demand also falls off away from the farm. A point is eventually reached where demand for beer at the delivered price, 60p in the example

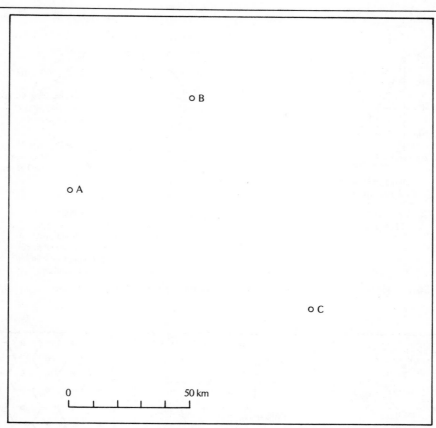

Figure 1.35   The location of three cement plants

Table 1.20   Charges for the three cement plants

Plant	Operating costs (£ per tonne of cement)	Loading charge (£ per tonne of cement)	Transport charge (£ per tonne of cement per kilometre)
A	5	1	0.10
B	5	3	0.10
C	10	2	0.10

(Figure 1.36a), is zero. At and beyond this distance from his farm, the farmer can sell no beer. Lösch showed that if the price axis is rotated through 90 degrees so as to lie horizontally, and if the graph is then turned full circle about the farm, a demand cone is produced (Figure 1.36b). The volume under this demand cone represents the total number of consumers of beer that the farmer can supply. The farmer's market area is thus, in theory, circular, the border corresponding to the distance at which the cost of carting the beer to consumers becomes so big that the delivered price is prohibitively large and consumers are unwilling to buy it (Figure 1.37a).

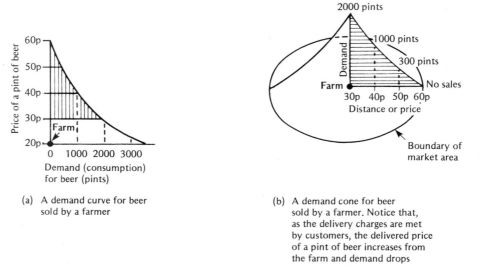

(a) A demand curve for beer sold by a farmer

(b) A demand cone for beer sold by a farmer. Notice that, as the delivery charges are met by customers, the delivered price of a pint of beer increases from the farm and demand drops

Figure 1.36   How Lösch defined the market area for a farmer producing beer

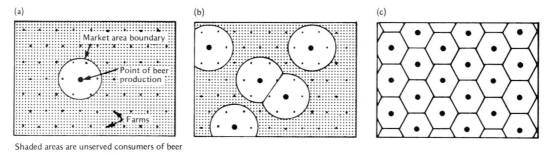

Shaded areas are unserved consumers of beer

Figure 1.37   Stages in the growth of a stable economic landscape according to August Lösch

Other farmers set up in the beer-producing business. As more and more farmers catch on, so the market areas become smaller and smaller (Figure 1.37b). Eventually, by the time the entire region is supplied with beer, the market areas are hexagonal in shape (Figure 1.37c). The hexagon provides the largest number of consumers for every producer and also provides the least delivery distance between a producer and his surrounding consumers.

THE BREWING INDUSTRY: A CASE STUDY

The brewing industry in the United Kingdom was originally dispersed and orientated towards local markets. This was because bulky, semi-perishable barrels of beer were distributed by horse and cart — a slow process. Also the basic ingredient in beer-making — water — could be found almost anywhere. The industry was characterized therefore by small, local breweries, the market areas of which would have followed the Löschian pattern. In the early nineteenth century, only London and Burton-upon-Trent had large breweries, London as a result of a large and dense urban market and Burton-upon-Trent because the water from the Trent river was ideal for making sparkling pale ale which enjoyed an international reputation.

Today, water (from the mains) is still the most important single item in beer-making, constituting 90 percent of the product. But there has been a reduction in the number of breweries, from 48,000 in 1839 to less than 150 in 1979. This change is the result of a series of takeovers and amalgamations. The industry is now, as it always has been, urban based, but tends to be found in larger settlements. Recently, with a rise in the demand for 'real ale', several small breweries using traditional methods have been set up, many of them serving a handful of pubs.

Brewing as an Urban-Based Activity

Figure 1.38 shows the relationship between employment in brewing and malting and population in the counties of England, Scotland, and Wales in 1976. It was constructed like this. The vertical axis represents the number of employees in the brewing and malting industry in each county expressed as a percentage of the total number of employees in brewing and malting in all counties. For instance, Cheshire employed 3.5 thousand people in brewing and malting. The total number of people employed in brewing and malting in all counties was 68.1 thousand. So, Cheshire's share of the national brewing and malting workforce was $(3.5/68.1) \times 100 = 5.14$ percent. The horizontal axis represents the population of each county expressed as a percentage of the national population. For instance, the population of Cheshire was 916.4 thousand. This figure is 1.74 percent of a national population (England, Scotland, and Wales) of 52,813 thousand. Both axes use a logarithmic scale. This simply spreads out the points and makes the relationship clearer.

If the proportion of the population employed in brewing and malting in each county were the same in all counties, all the points plotted in Figure 1.38 would lie on the 45-degree line; the relative proportion of the population of a county employed in brewing and malting would remain constant, even though in absolute terms more populous counties would have a bigger brewing and malting workforce.

1a Draw a **line of best fit** through the data points on Figure 1.38. This line must pass through the point of intersection of the mean county population expressed as percentage of the national population (1.52 percent) and the mean number of employees in brewing and malting expressed as a percentage of the national employment in brewing and malting (also 1.52 percent).

 b On the not unreasonable assumption that the more populous regions have bigger urban populations, what can you infer about the regional distribution of brewing and malting employment in England, Scotland, and Wales? (To answer this question, you will need to study Figure 1.38 and decide whether or not more populous counties have a bigger or lesser proportion of employees in brewing and malting and whether or not less populous counties have a bigger or lesser proportion of employees in brewing and malting. Remember that a county with a fair share, no more or less, will lie on the 45-degree line.)

Rationalization of the Brewing Industry

Improvements in transport technology have enabled breweries to serve large market areas. Larger breweries, because they have less overheads, can produce a pint of beer cheaper than smaller breweries can — this follows the principle of internal economies of scale. With lower production costs, larger breweries can command bigger market areas than smaller breweries.

We saw how this worked in theory on p. 52. The increase in the market areas of breweries has led to rises in distribution costs, but until recently these rises have been more than compensated by savings in production costs.

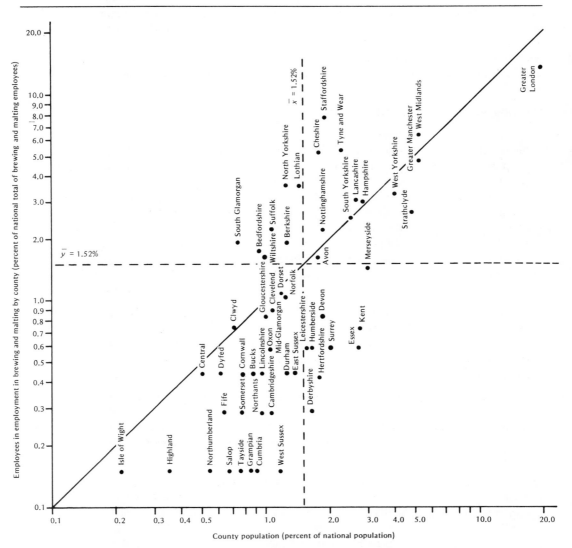

Figure 1.38   The relationship between employees in brewing and malting and population. The 'nation' in this instance means England, Scotland, and Wales. The following counties had no employees in employment in brewing and malting in 1976: Herefordshire, Warwickshire, Gwent, Gwynedd, Powys, West Glamorgan, Borders, Dumfries and Galloway, Orkney, Shetland, and Western Isles.
Source of data: Brewing and malting employment — Census of Employment, 1976. Kindly supplied by the Department of Employment, Orphanage Road, Watford. Population — *Geographical Digest*.

The relationship between distribution costs, production costs, and economies of scale can be expressed in what is called a **substitution curve** or **opportunity cost curve**. Figure 1.39 is an hypothetical substitution curve for a brewery. Production costs are given on the horizontal axis and distribution costs are given on the vertical axis. At the point in time shown, the brewery has unit production costs of £17.50 and unit distribution costs of £5.00. Its total costs are thus £22.50.

2 If the brewery decides to increase production by increasing the output of one of its plants and closing down others, its production costs will be reduced but distribution costs will increase. At which point along the substitution curve would the savings in production costs be offset by increases in distribution costs? (To answer this question, you will need to compute total costs at sample points along the substitution curve.)

The Whitbread brewery has pursued a policy of reducing the numbers of plants and increasing production in the remaining ones. In the period 1960 to 1971 the company closed fifteen of its plants. But the rationalization of the brewing industry, as the process of shutting down some plants and expanding others is known, seems to have passed its peak. The dramatic increase in the cost of oil has increased both distribution and production costs, more so in larger plants than smaller ones because the large plants which produce processed beers have greater energy requirements per unit of output than small plants running on traditional methods. So, since 1974, market area size and scale of production have contracted.

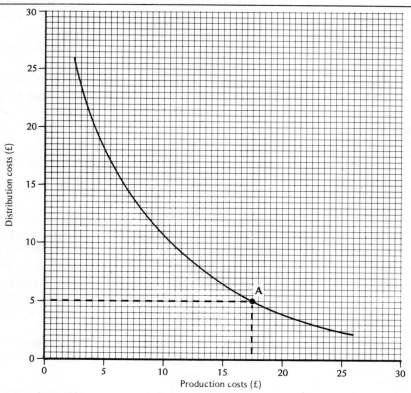

Figure 1.39   A hypothetical substitution curve for a brewery

Cost and Revenue

The theories of industrial location we have looked at fall roughly into two schools. The first, or **least-cost, school** seeks to find the site of least cost. It has at its roots the work of Weber and includes much of the work of Palander and Hoover. Its chief limitation is the assumption of demand being constant and everywhere the same. If demand is allowed to vary, as in the real world it does, then the site of least costs may not necessarily yield the biggest profits — extra sales at another site may more than compensate for higher costs there. The second, or **market-area, school** takes the opposite view to the first in assuming that production costs are the same everywhere and looking at the relationship between competition between firms and market areas. To this school belongs some of the work of Palander and Hoover, much of the work of Lösch, and the work of economists like Hotelling. Its chief limitation is the assumption that production costs are everywhere the same.

A third school of theories attempts to bring together the other two schools by allowing production costs and demand to vary at the same time. The first attempt at a synthesis of the 'least-cost' and 'market-area' schools was made by Melvin Greenhut in 1956 in a book called *Plant Location in Theory and in Practice*. Greenhut's theory of industrial location includes a consideration of several factors. First of all, it considers cost factors of location — procurement costs, distribution costs, labour costs, processing costs, capital, and taxation; these are essentially the factors considered by the 'least-cost' school. Secondly, it considers demand factors of location, the attempts of firms to monopolize parts of a market; this is the facet of location studied by the 'market-area' school. It then includes consideration of factors which reduce costs, such as agglomeration and deglomeration; factors which increase revenue (agglomeration and deglomeration will again serve as an example); and personal factors, such as good relations with suppliers and bankers, which tend to reduce costs, good existing contacts with customers, which tend to increase revenue, and an entrepreneur's liking for an area.

D. M. SMITH'S THEORY OF INDUSTRIAL LOCATION

In a paper published in 1966, British geographer David Smith put forward a theoretical framework for geographical studies of industrial location. In 1971 Smith restated and expanded this framework in a book called *Industrial Location*. Smith's theory draws heavily upon earlier theories but is more flexible than its predecessors, allowing demand, for example, to vary, and provides a better guide for interpreting and analysing actual patterns of industrial location.

The Location of a Bakery

The starting point of Smith's thesis is the assumption that an individual or a group who set up a business — let's say Mr. Dough who wishes to set up a bakery — will be out to make money. Moreover, though this constraint can be relaxed later, he will be out to make as much money as possible. So assuming, the bakery will be located at the place that brings in the biggest profits; at least it will if Mr. Dough has the knowledge and ability to find this fortune-making place.

The cost of producing bread will vary from place to place according to the cost of procuring raw materials — flour, sugar, salt, yeast, and the many additives; the cost of making bread from the raw materials in the production process; and the cost of distributing bread to the market. The revenue obtained from selling bread varies from place to place according to demand for bread which itself depends partly on the location and strategy of competitors. The site with the biggest profit is found where revenue exceeds costs by the largest amount.

Figure 1.40 shows some possible variations in average costs of producing bread and the price of bread. The average cost of producing a loaf of bread is lowest at point A and increases in either direction away from point A. Thus it is cheapest to procure materials and make bread at point A. The price of a loaf of bread is highest at point B, where demand is highest, and falls off in either direction away from point B. (Smith called the line showing the average cost of a unit of production, in our case a loaf of bread, a **space-cost curve**, and the line showing the price of a unit of production, again a loaf of bread, a **space-revenue curve**.) At any point on Figure 1.40, the difference between the price and the average cost of a loaf of bread represents the average profit or loss made on one loaf. Where the price of a loaf is more than the average cost of making it, Mr. Dough will make a profit. The size of the profit can be found for any point by subtracting the average cost from the profit. Where the price of a loaf is less than the average amount it cost to produce, Mr. Dough will make a loss. The size of the loss can be found for any point by subtracting the price of the loaf from the average cost of making it. Notice that from this information, the area in which it will be profitable for Mr. Dough to make and sell bread can be delimited. The boundaries of this area, or what Smith termed **spatial margins of profitability**, correspond to points where average cost and price are the same. At these points Mr. Dough will just break even.

Smith argued that, as a general principle of industrial location, the following statements are true:

i Variations in cost and revenue from one place to another produce a location where profits are biggest.

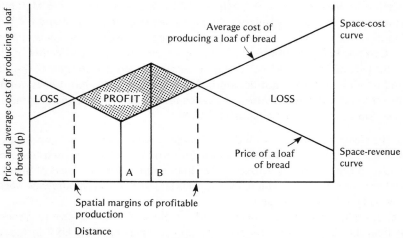

Figure 1.40   D. M. Smith's space-revenue and space-cost curves illustrated for Mr. Dough's bakery

ii Variations in cost and revenue from one place to another produce spatial margins beyond which a firm cannot operate and expect to make a profit.

iii Within spatial margins, a firm is free to locate anywhere providing that the site yielding the greatest profit is not sought. This is a relaxation of the assumption that an entrepreneur will be out to make as much money as possible; rather he will settle for a satisfactory living (compare with the discussion on p. 68).

1 Mark on Figure 1.41 areas of profit, areas of loss, and spatial margins of profitability.

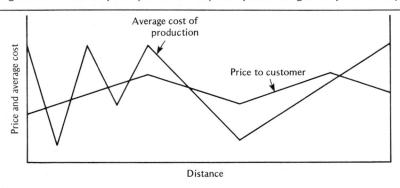

Figure 1.41

Electronic Equipment Manufacture in the United States

To demonstrate how his framework could be used in practice, Smith drew up cost surfaces for a number of industries in the USA. These surfaces describe variations in total cost and not just transport cost. An example is the cost surface for the manufacture of electronic equipment (Figure 1.42a). The actual location of the electronic equipment industry in 1963 is shown in Figure 1.42b. Areas of low cost in Figure 1.42a do not correspond with areas producing electronic equipment as shown in Figure 1.42b. The main area of production concentration, a belt extending from Massachusetts, through New York and New Jersey, into Maryland, is an area of varying cost levels. Two secondary production centres, the Midwest and the West Coast, correspond to the areas of highest operating costs. In the low-cost zone which occupies the South, South Carolina being the cheapest place, there is virtually no manufacture of electronic equipment. This lack of correspondence certainly means that the distribution of the electronics equipment industry in the USA is not determined by operating costs; but it does not mean that Smith's approach is invalid. Other factors, not taken into account in the analysis, may play dominant roles. Smith thought these factors might be:

i Labour in the electronic equipment industry is highly skilled and areas of existing concentration would have workers of the right kind — a new firm would be better advised from this point of view to locate in Boston or Los Angeles rather than South Carolina. Areas of high labour costs may reflect greater skill and productivity and thus labour costs given purely as wage rates cannot always be taken at face value.

ii Economics of agglomeration may come into play making high labour cost locations more attractive than they appear. The electronic equipment industry in the USA includes sophisticated apparatus for missile systems and space exploration and is associated with scientific research and development. Electronics firms have much to gain from close contact with other plants and universities engaged in related activities. This degree of clustering is particularly evident in California where the industry currently employs 25 percent of the state's manufacturing labour force. Moreover, four metropolitan areas — Anaheim-Santa Ana, Los Angeles, Long Beach, San José and San Diego — house 82 percent of all electronics firms.

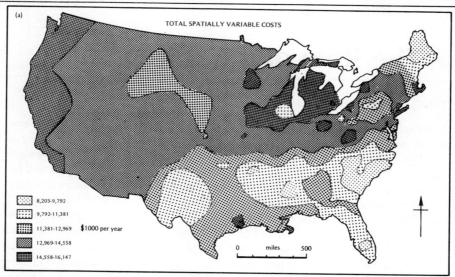

Figure 1.42a   The manufacture of electronic equipment: the total-cost surface
Source of data: The Fantus Company
Reprinted with permission from *Industrial Location* by D. M. Smith (1971), published by John Wiley & Sons, New York, figure 18.17

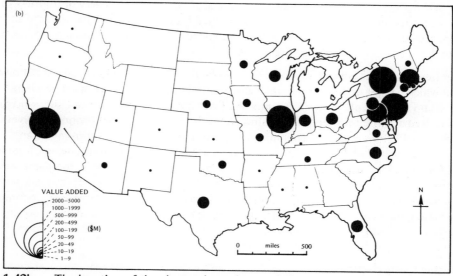

Figure 1.42b   The location of the electronic equipment industry, 1963
Source: Census of Manufacturing, 1963
Reprinted with permission from *Industrial Location* by D. M. Smith (1971), published by John Wiley & Sons, New York, figure 18.18

The Role of the Entrepreneur

SELECTING A SITE

In seeking to explain the location of industry, geographers have tended to look at economic factors, in particular raw materials, power, labour, markets, and agglomeration, weighing up the effects of each in a given case. Most of the theories of industrial location that we have studied deal with these economic factors. They assume that the entrepreneur is all-knowing and behaves in a rational and objective way. But this is not usually how decisions about where to locate industries are made. Economic factors are important but so too are the whims and fancies of the entrepreneur. This applies where the entrepreneur is an individual and where the entrepreneur is the board of a company. If an entrepreneur knew the expenditure and revenue for all sites in an area, he could select the site which would bring in the biggest profits. But this is not usually possible. Seldom does an entrepreneur have at his disposal enough information to find the most profitable site. Indeed, he may not wish to make as much money as possible. He is not a so-called **economic man** who acts in a rational and objective way. More likely than not, armed with incomplete information, he will either try to come to a rational decision about where to set up his business or he will simply establish a broad area in which it seems feasible to run his business and then select a site according to seemingly irrational criteria such as scenery or nearness to a golf course. Under these circumstances, the site selected is unlikely to coincide with the site of maximum profit. Nonetheless, the entrepreneur will be happy with his potential returns as he sees them. He is a so-called **satisficer** (Figure 1.43).

Not all entrepreneurs act alike. A survey conducted by P. M. Townroe (1971) among 59 firms in Britain which had set up branches or relocated showed that the basis for deciding where to move or set up a branch plant varied enormously. Five firms considered just one site, whereas 24 firms looked at more than ten possible sites. Some 20 firms carried out economic surveys of several sites, 13 firms made no economic surveys, and the remaining 26 firms made an economic evaluation of only the site chosen. If any pattern did emerge from Townroe's survey, it was as follows. Small firms tend to know of a site by chance, search a small, local area, have many personal contacts, and consider a few possible sites. In short, they adopt the behaviour of a satisficer. On the other hand, large firms search systematically over a large area, commonly over the nation, have contacts with official agencies, and give consideration to many possible sites. In short, they adopt the behaviour of an economic man or optimizer. But of course, 59 firms is a small sample from which to make general statements.

Another study, which reveals the range of factors influencing an entrepreneur in his deciding where to locate a plant, was made in 1964. In this, the McGraw-Hill plant site survey, questionnaires were sent out to business and industrial subscribers of a leading US business journal. The firms were asked to list the factors they deemed important in deciding on a site

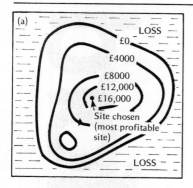 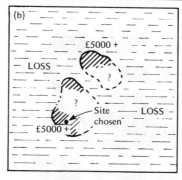 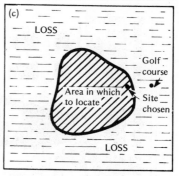

Figure 1.43   Three different backgrounds against which decisions about where to locate an industry are made

(a) The expenditure and revenue for all possible sites are known so the potential profits at all sites can be mapped. The site selected is the one which offers the largest potential profit. It has been chosen by an entrepreneur acting in a rational and objective way with all the necessary information at his disposal. This kind of entrepreneur is called an economic man or optimizer.

(b) The information about potential profits is patchy. The entrepreneur, in choosing a site, was happy with a potential profit of £5000 and, armed with his incomplete information, has selected a site with that potential profit. He probably realizes that bigger profits could be obtained elsewhere but he is satisfied with the site he chooses. He is called a satisficer.

(c) The entrepreneur has simply established an area in which it seems feasible to locate his industry. Within this area he assumed all sites were equally attractive in terms of profit and his selection of a site has been swayed by personal factors — nearness to a golf course for instance. He too is a satisficer.

for a new plant. As many as three-quarters of the respondents indicated that good transport facilities were important. Between one-half and three-quarters of the respondents felt that these factors were important: reasonable cost of property, reasonable or low taxes, ample area for expansion, favourable labour climate, favourable attitude of community and residents to industry, nearness to present sales area, reasonable cost of construction, and favourable climate for personnel. Between one-third and one-half of the respondents listed these factors: availability of labour skills, access to utilities, near sources of raw materials, need for plant to service new or expanding sales area, favourable political climate to business, pleasant living conditions, commercial facilities, rail facilities, zoning restrictions, cost of living and economic conditions, and labour rates. Less than one-third of the respondents mentioned these factors: educational facilities, favourable climate for productive process, inexpensive fuel and power, public transport, recreational and cultural facilities, water supply, waste disposal, near airport, and topography.

1a Study the advertisements produced by Telford Development Corporation and Peterborough Planning Department (Figure 1.44). List the things they seem to think important in attracting new businesses.

 b In what ways is the emphasis different in the two advertisements?

70

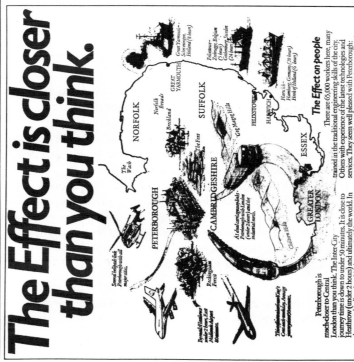

Figure 1.44 Recent advertisements designed to attract industry to Peterborough and Telford

THE BEHAVIOURAL SCHOOL

Attempts have been made to clarify the role of the entrepreneur as a satisficer in industrial location by members of the so-called **behavioural school**. Allan Pred, in his book *Behaviour and Location*, examined the factors which influence how an entrepreneur behaves in selecting a site. He recognized two key factors:

i   the amount of information an entrepreneur has to hand (Pred called this the perceived information); and

ii the ability an entrepreneur has to use this information.

He devised a 'behavioural matrix', the two axes of which measure perceived information and the ability to use information (Figure 1.45a). Different parts of the matrix represent different behavioural patterns in entrepreneurs. These differences will influence the site selected to set up a firm within spatial margins of profitability. An 'able' entrepreneur with access to a lot of information is likely to locate his firm near the optimum economic location whereas a 'less able' entrepreneur with access to limited information stands a good chance of locating his firm near the margin of profitable production (Figure 1.45a).

In a similar vein, Eliot Hurst devised a 'managerial matrix' (Figure 1.45b), the axes of which are:

i   an entrepreneur's concern for profit; and

ii an entrepreneur's concern for other utilities.

Different parts of this matrix represent different sets of entrepreneurial concern. These differences will influence site selection (Figure 1.45b).

The selection of a site when both the behavioural matrix and the managerial matrix are considered together may be a complex affair. Figure 1.45c depicts the way in which different combinations of behavioural and managerial facets can lead to the selection of different sites by entrepreneurs in the same industry.

THE VAGARIES OF ENTREPRENEURIAL DECISIONS

The inclusion of personal factors of location in Greenhut's and Smith's theories marked the fall of economic man, the hero of the least-cost and market-area schools, and the rise of satisficer man, whose role in industrial location is now recognized as being of paramount importance. Certainly, the whims and fancies of an entrepreneur can be a stumbling-block in the application of 'classical' locational theory, as the following examples will show.

The Eighteenth-Century Copperworks in Macclesfield

Charles Roe, an entrepreneur born in 1715, decided in 1758, for reasons best known to himself, to set up a copperworks in Macclesfield, Cheshire (Figure 1.46). Admittedly, coal was available on nearby Macclesfield Common and copper ore was available at Alderley Edge some seven miles away; but neither of these sources lasted for long. Coal had to be hauled eight miles over land from Poynton. Copper ore was mined at Alderley Edge for about ten years but supplies dwindled and two other sources were used. Copper mined at Coniston in the Lake District was carried by horses to Morecambe Bay and shipped to the Mersey estuary. From there it was taken by barge down the Weaver Navigation to Northwich whence it was taken overland to Macclesfield (and two other copperworks owned by Roe at Havannah, near Congleton, and at Bosley, just south of Macclesfield). Later, copper mined in Anglesey was brought to Macclesfield by a less-complicated

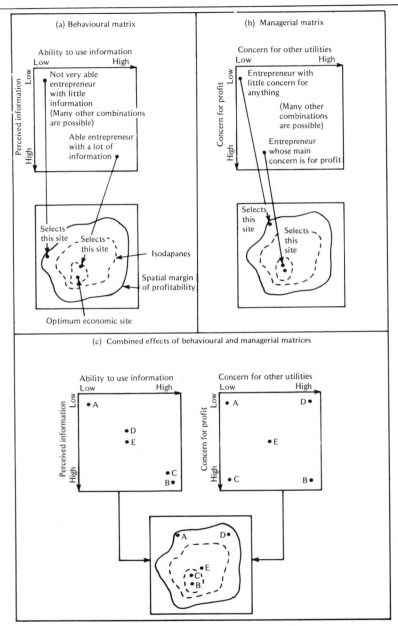

Figure 1.45 The letters on (c) correspond to different types of management (entrepreneurs). A is 'impoverished management' — little motivation or commercial ability. B is 'team management' — complete motivation and commercial skills. C is 'task management' — high profit motive and high scores on behavioural matrix. D is 'country club management' — low profit motive (e.g., monopoly situation). E is 'dampened pendulum' — a balance between economic survival and satisfaction, together with average behavioural attributes. (The management 'types' are after Eliot Hurst, 1972)

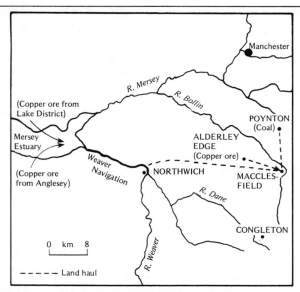

Figure 1.46   Supply routes of materials for Charles Roe's eighteenth-century copperworks in Macclesfield, Cheshire

route than that from Coniston but the haul was still difficult, long, and costly. To cut costs, Charles Roe tried to have a canal built between the River Weaver and Macclesfield and between Macclesfield and Poynton, canal transport at that time being the cheapest and least troublesome mode of moving heavy materials. But the plan was stopped by the Duke of Bridgewater, the great canal builder, who successfully petitioned against it, probably, it has been suggested, to secure a more extensive and lucrative monopoly for himself. In 1767, a year after the failure of the canal scheme, Charles Roe and partners, in order that they might reduce transport costs, set up a copper-smelting plant on the Mersey estuary near Liverpool. The extracted copper was then sent from Liverpool to the Macclesfield area. Charles Roe died in 1781. Twenty years later the Macclesfield copperworks had vanished, the only remaining signs of its existence being streets bearing names such as Copper Street and Smelthouses.

1   What features of the eighteenth-century Macclesfield copperworks can be explained by theories of industrial location?

The local sources of copper and coal having run out, the copperworks in Macclesfield was, perhaps surprisingly, not shut down. Alternative sources of raw materials were produced with considerable difficulty and the copperworks survived. This is an example of **geographical inertia,** the tendency for an industry to remain at a particular site even though other sites are more profitable.

Automobile Manufacturing: a Footloose Industry

The whimsical behaviour of the entrepreneur in deciding where to locate an industry is commonly prevalent in footloose industries. Footloose industries have an exceptional degree of locational freedom (wide spatial margins of profitability) because their material inputs tend to be high-value, semi-finished or finished goods (components); their markets may be local, national, or international; and their transport costs are a negligible portion of their total costs.

74

The automobile industry is the classic example of a footloose industry and it also illustrates how the behaviour of individuals may influence location. R. L. Morrill argued that, if the automobile industry were to start afresh in the United States, it would probably set up shop in Chicago. Detroit and its environs were the place where entrepreneurial and technological expertise happened to arise in the form of Henry Ford, the Dodge brothers, and M. Leyland, the founder of Cadillac cars.

2 Today, the lower Great Lakes region produces just one-third of the US output of automobiles. Why do you think the industry has spread to the other areas shown in Figure 1.47a? (Bear in mind the information mapped in Figures 1.47b, c, and d.)

An Exercise in Decision-Making

Any location within the spatial margins of profitability of an industry can be regarded as satisfactory, but only one (or perhaps a few) sites will bring in the biggest profits.

3 On Figure 1.48 are indicated the spatial margins of profitability for sugar-beet processing in England and Wales. If you were given the opportunity of setting up a sugar-beet processing factory, where would you site it? Give reasons for your choice. (Bear in mind all you have learned about factors of industrial location, especially the importance of costs and sales. You do not know where the biggest profits from sugar-beet-processing are to be made and personal considerations may come into play — perhaps, for instance, you have an aversion to Norfolk. It might be interesting to build up a composite map showing the sites for the sugar-beet factory selected by all members of your class and to compare notes as to what factors influenced the various locational decisions that were made.)

Figure 1.47a   Automobile assembly plants in the United States, 1966
Reprinted with permission from *Industrial Geography* by R. C. Riley (1973), published by Chatto and Windus, London, figure 79

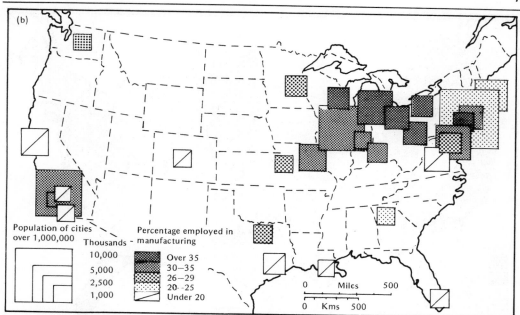

Figure 1.47b Percentage employed in manufacturing in the United States, 1969
Reprinted with permission from *A Modern Geography of the United States: Aspects of Life and Economy* by R. Estall (1972), published by Penguin Books, Harmondsworth, figure 42

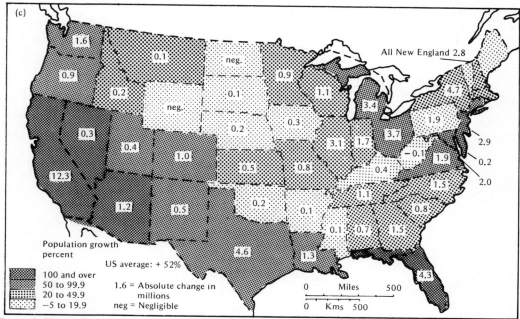

Figure 1.47c Population change in the United States, 1940 to 1968
Reprinted with permission from *A Modern Geography of the United States: Aspects of Life and Economy* by R. Estall (1972), published by Penguin Books, Harmondsworth, figure 4

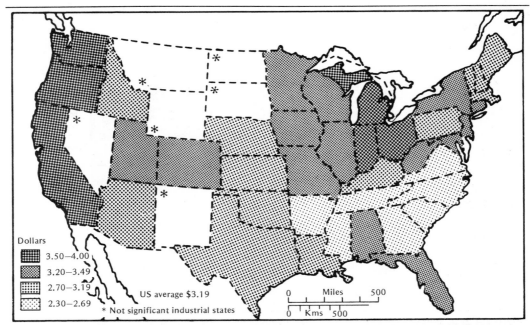

Figure 1.47d   Average hourly earnings of production workers in manufacturing, 1969
Reprinted with permission from *A Modern Geography of the United States: Aspects of Life and Economy* by R. Estall (1972), published by Penguin Books, Harmondsworth, figure 43

Figure 1.48   Where would you locate a sugar-beet factory? The shaded areas lie within the approximate spatial margins of profitability for sugar-beet production.

Summary

The location of an industry, be it extractive, manufacturing, or service, is influenced by many and various factors including raw materials, fuel supply, labour supply, markets, and, by no means least, the whims of entrepreneurs. Theories which attempt to explain the location of industry, because they must of necessity simplify the complex, real-world situation, hold some factors constant and allow others to vary. Three main groups of theories may be recognized. Theories in the first group, which includes the work of Weber and part of the work of Hoover and Palander, focus on the role of variations in cost from one place to another. Weber's chief concern was with finding the site with the lowest transport costs; but he also considered the effect of cheap labour and the effect of agglomeration. Hoover and Palander refined Weber's ideas, looking in more detail at the structure of transport costs. All of these 'classical' cost theories have limited applications to industries in the modern world.

Theories in the second group, which includes part of the work of Hoover and Palander, hold costs constant and look at the effect of variations in revenue (demand) from one place to another. In these theories, the role of the market is the main concern and market areas between competing businesses may be predicted. Probably the most comprehensive theory in this group, and perhaps the most difficult to follow, is that put forward by August Lösch to explain the structure of an economic landscape.

Theories in the third group, which includes the work of Greenhut and of Smith, look at cost and revenue varying from place to place at the same time. They also take into account the effect that entrepreneurial decisions may have on industrial location. Because they make fewer simplifying assumptions, these theories tend to have wider applicability than those in the other two groups.

Further Reading

Location in Space, P. Lloyd and P. Dicken, Harper & Row (1977).
Industrial Geography, R. C. Riley, Chatto and Windus (1973), Chapter 1.
The Location of Manufacturing Industry, J. Bale, Oliver & Boyd (1977), Chapters 5 and 6.
Human Geography: Theories and their Applications, M. G. Bradford and W. A. Kent, Oxford University Press (1977), Chapter 3.

CHAPTER TWO
THE PATTERN OF INDUSTRY

The first chapter dealt with the location of individual industries. This chapter will look at the pattern of industry as a whole. In particular, we shall explore the following questions: Why should industry tend to concentrate and thrive in some areas while in other areas it is thin on the ground and those industries that there are seem unable to make a go of it? Why should well-off cores of industrial growth surrounded by relatively poor peripheries of industrial stagnation and decline be found in cities, in nations, and even over areas of international extent?

Theories of Regional Growth and Decline

PROCESSES OF REGIONAL CHANGE

Several theories, all rather descriptive in nature, have been put forward to explain why some regions thrive while others stagnate. The key seems to lie in the role core regions play in invention and innovation. A core region is a seed-bed for new developments coming from the fertile minds of highly skilled scientists, technologists, and managers. These new developments, through a chain of events, attract industries from the periphery to the core. The core grows bigger, further industrial developments take place, and even more industries move in. The process is self-generating. It acts in a circular and cumulative fashion, the core region growing during each cycle of expansion. The cycle of events has been studied in detail by the Swedish economist Gunnar Myrdal who gave it the name **cumulative causation**. An illustration of a typical cycle is shown in Figure 2.1.

Consider what happens when a new industry is set up in a core region. To start with, new jobs are generated, people move in to take them, and the purchasing power of the population grows. This increases the demand for houses, schools, consumer goods, and services, so creating even more jobs. The new industry will also attract other industries which supply it with raw materials or use its products. In other words, it triggers small-scale agglomeration. This creates extra jobs in services, public utilities, and construction. Yet more industries are then attracted to the area by the larger labour pool, a bigger local market, and better developed back-up services in the expanded core region. So the establishment of just one new industry in the core region can, by a complex chain of events, boost the local economy in unexpected ways. This is known as the **regional multiplier effect**. Figure 2.2 shows how the effect of a new steel plant in the New York-Philadelphia region goes far beyond the direct creation of jobs at the plant itself. New employment is generated in many types of manufacturing and service industries. This gives increased scope for new developments in the steel plant and increases the regional population, both of which pave the way to the expansion of the steel plant. And so the cycle goes on.

A growing core region feeds on resources of surrounding areas, tapping people, capital, and goods. These **backwash effects**, as Myrdal called them, may be harmful to the demographic and economic structure of the peripheral supply regions. For instance, the movement of people is selective, younger people being attracted to the core area. This leads to a preponderance of older folk in the peripheral areas.

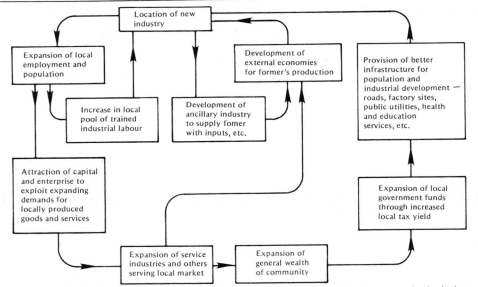

Figure 2.1   Myrdal's process of cumulative causation. (The two subcircles are included to illustrate the ramifications of the process.)
Reprinted with permission from 'Models of economic development' by D. E. Keeble (1967) in *Models in Geography*, edited by R. J. Chorley and P. Haggett, published by Methuen, London, figure 8.4

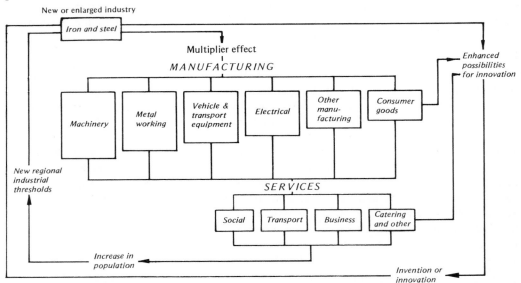

Figure 2.2   The multiplier effects of the location of an integrated steel plant in the New York—Philadelphia area. The boxes are drawn in proportion to the increase in employment that was thought likely to occur after six cycles of expansion.
Reprinted with permission from 'Models of industrial location' by F. E. Ian Hamilton (1967) in *Models in Geography*, edited by R. J. Chorley and P. Haggett, published by Methuen, London, figure 10.15

1. Complete Figure 2.3, which shows a vicious circle of labour migration, by inserting the appropriate labels i to iv in the empty boxes.

The core region attracts people who live nearby: there is a lack of long-distance migration. This is partly accounted for by Samuel Stouffer's **theory of intervening opportunities** which states that migration of people between two regions is curtailed by the number of opportunities (towns and jobs) on the route between them. There are likely to be fewer intervening opportunities between a core and neighbouring regions than between the core and far-away places.

Backwash effects are unlikely to drain peripheral regions of all resources. As Myrdal explained, the thriving core region has an ever-increasing demand for raw materials and agricultural products. But new developments in the core, including special machinery, fertilizers, and hybrid crops, make an increase in food production possible. The periphery may also benefit from the spread of consumer goods and branch firms from the core. Myrdal called these processes, by which growth in the periphery is stimulated, **spread effects**.

Spread effects operate, in part, through **growth poles**. The basic idea of a growth pole, as developed by the French economist François Perroux, is that a new site of expanding industries, usually in a city, will set off chain reactions of industrial and economic development throughout the hinterland. But J. R. Boudeville pointed out that, for a city to have this impact, its industries must be of the propulsive type, that is, they must exert a direct and indirect influence over most other activities

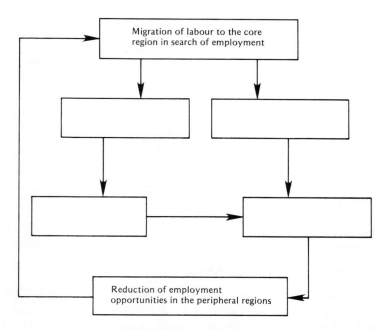

i   Decrease in attraction of peripheral regions to new industry
ii Smaller local market and reduced purchasing power
iii Decline in local services
iv Change in age structure of the population

Figure 2.3   A 'vicious circle' of labour migration from peripheral regions to the core region

by intense linkage with other industries and the rest of the economy. Only then will the desired multiplier effects come into play. Propulsive industries are generally those which have the greatest capacity for invention and innovation. This usually means large, flourishing firms for they alone have the capital to invest in extensive research and development programmes.

Growth poles have been set up to try to stimulate industrial and economic growth in some peripheral regions. But to have a chance of achieving self-generating growth, a growth pole has to possess something in the order of 150,000 to 250,000 inhabitants. Investments in centres in problem areas of the United States during the 1960s were spread too thinly among centres of greatest need, rather than those with the greatest potential growth. The result was that the would-be growth poles simply soaked up all the investments poured into them without growing. The problem with growth pole policies is finding the best potential site for growth and then convincing the populace of rejected sites of the wisdom of the policy. In 1970 it was announced that Catanzaro in the Italian Mezzogiorno was to be a regional centre. Riots broke out in Reggio de Calabria, the residents of which town clearly thought they had a stronger case. Growth pole policies have been used by French planners who in 1961 established *métropoles d'équilibre*, including Bordeaux and Strasbourg, to counterbalance the economic dominance of the Paris region (Figure 2.4). These *métropoles* were awarded substantial government aid to help attract industry and thereby stimulate the local economy.

Figure 2.4   *Métropoles d'équilibre*

THE PATTERN OF REGIONAL CHANGE

All processes of regional development have been put together by American planner John Friedmann and presented as a regional development model. Four main types of region are recognized in the model: core regions, upwards transition regions, resource frontier regions, and downwards transition regions. **Core regions** are centres of growth in which new industry and innovations, such as the development of fertilizers and hybrid crops, proliferate. They can be recognized at various scales ranging from international cores (northeastern France, the Ruhr, and Benelux), through national cores (London and Birmingham), to regional and local centres. **Upwards transition regions** are located outside core regions but, usually because they are well endowed with natural resources, are areas of immigration and increasing investment and agricultural production. The immigrants tend to settle in many small centres rather than one big one. **Development corridors** are a special form of upwards transition region. They lie between two core cities. The Rio de Janiero-São Paulo corridor in Brazil is an example. **Resource frontier regions** also lie outside core regions. They are zones of new settlement where virgin territory is occupied and made productive. A good example is the temperate grasslands during their colonization phase in the nineteenth century. Agricultural colonization on such a vast scale is very limited today but the Soviet occupation of the virgin lands of Siberia is an instance of it. Most present-day resource frontier regions are associated with mineral exploitation, as in northern Alaska, and commercial forestry. In Britain, the area around Aberdeen, owing to the exploitation of North Sea oil, shows signs of becoming a resource frontier region. **Downwards transition regions** are peripheral areas of old, established settlements whose economic fortunes are on the wane. Signs of economic stagnation and decadence abound, a result of the depletion of mineral resources or the ageing of industrial complexes. Their rural economies are also sluggish or even on the decline. There is little incentive for people to stay so they move

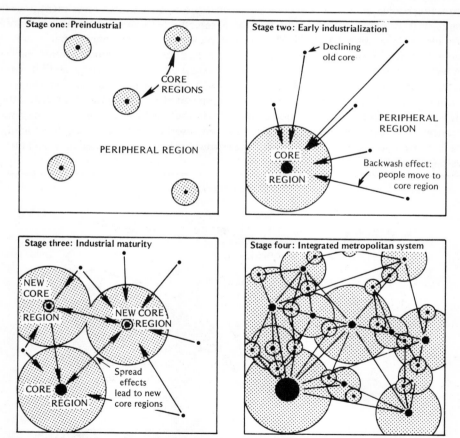

Figure 2.5   An interpretation of Friedmann's model of economic development
Based on ideas in *Regional Development Policy: A Case Study of Venezuela* by J. Friedmann, published by MIT Press, Cambridge, Massachussetts

84

away. Examples include the Italian Mezzogiorno, old industrial centres in the United Kingdom — the North East, North West, and South Wales, and ghetto areas in cities. In fact, the decline of the regions in the United Kingdom has been partly averted by government assistance, more truly downwards transition regions being mid-Wales and highland Scotland.

Friedmann described how the regions grow through four stages (Figure 2.5). He identified these stages, in broad terms, in the regional development of Venezuela (Figure 2.6). In stage one, small and independent core regions are found, no one core being dominant. In Venezuela, this stage seems to have been achieved by 1936 when there were a number of fairly distinct regional economies centred on Maracaibo, San Cristobal, Ciudad Bolivar, and a few other towns, all of which were loosely governed from the capital city, Caracas. In stage two, during early industrialization, one or two core regions come to dominate, possibly because they have more natural resources (for example, coal), lie in densely populated areas and so have a large market, or are favourably sited for trade with foreign markets (for instance, ports). The process of cumulative causation begins to take place, backwash effects drawing people into the core regions. In Venezuela, the Caracas region, being endowed with oil resources, emerged as the single national core, attracting people, industrial and commercial enterprises, government investment, foreign immigrants, and consumer industries. In stage three, spread effects come into play, partly offsetting backwash effects in peripheral regions. Thus, favourable parts of the periphery, that is, areas well endowed with natural resources, areas with large markets, or areas with good climates, may be transformed into new core regions. In Venezuela, this stage can be seen in the growth of secondary cores in the periphery at Maracaibo, Barquisimeto, and Valencia which accompanied the intensification of oil development. At the same time, spread effects from Caracas reached the Valencia basin where industrial expansion took place. In the fourth stage, spread and backwash effects cancel each other out and peripheral regions are absorbed into a complex, integrated, and efficient metropolitan system. This is an ideal stage and is perhaps approached in small nations of the developed world.

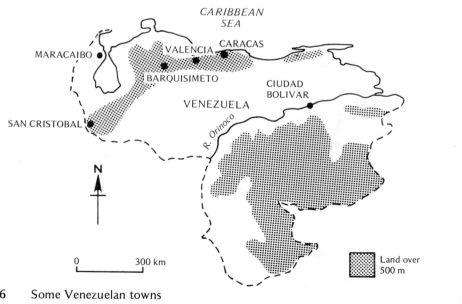

Figure 2.6   Some Venezuelan towns

Industry in the United Kingdom

Industry is spread unevenly around the regions of the United Kingdom. Phrases like 'the drift to the South' and 'the North begins at Watford' reflect the popular view of British industry that industries in the southern regions are flourishing while those in northern regions are on the decline. But postwar industrial growth in the United Kingdom should be seen, not as a conflict between North and South, but between a **core** area of industrial expansion and a **periphery** of less fortunate regions in which industry is sluggish or on the decline.

THE INDUSTRIAL CORE

The flourishing industrial core of the United Kingdom is centred on the large urban industrial complexes of London and Birmingham (Figure 2.7a). It stretches out to include much of the South East, the East Midlands, the West Midlands, and the Bristol area. The reasons it flourishes are many and varied. It has a large, fast-growing population; it has a supply of labour of the right quantity and of the right quality. In particular, the concentration of highly skilled research scientists, technologists, and managers in the South East region is of immense importance because it is the centre of industrial invention and innovation. Also, the industrial core seems to provide entrepreneurs with the right environment for setting up small firms to exploit new markets, techniques, or products. The birth-rate of small, 'back-room' firms is reputedly high in London and the West Midlands and this is undoubtedly linked with the pool of entrepreneurs in these areas — skilled workers keen to go it alone, scientists or technologists with new ideas, or foreign refugees with imported skills. Other factors which encourage new enterprise in the industrial core are the many small factory buildings available for rent or purchase, interest rates on industrial development as much as 2 percent lower than in other regions, and quick and easy access to customers (Figures 2.7b and c). Industries in the core area also benefit from external economies. There is ready contact with customers and with other industries supplying components. Specialist ancillary services are found which are absent from peripheral regions. Discussions with like-minded entrepreneurs, and the pooling of information and experience, are at a premium. In the light of all these advantages, it is no surprise that in the industrial core is concentrated most of the United Kingdom's manufacturing industry (Figure 2.8).

THE INDUSTRIAL PERIPHERY

The periphery of sluggish and declining industrial development in the United Kingdom surrounds the industrial core. It includes Wales, the North East, the Lancashire textile towns, Scotland, Northern Ireland, coastal parts of East Anglia and eastern Kent, and Devon and Cornwall. By and large, it lacks the demographic, economic, and locational advantages of the core region. It has a small, and

86

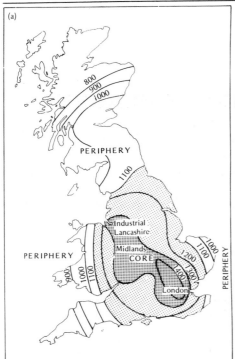

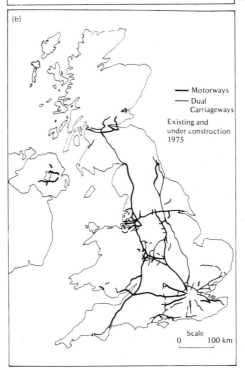

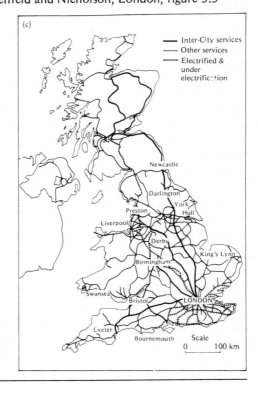

Figure 2.7a A map of economic (income) potential in England, Scotland, and Wales in 1966.
Economic or income potential is a measure of the relative nearness of a place to a national or international market. It reflects the variations in purchasing power from one place to another.
The parts of the map with a high economic potential have a greater power to attract economic activity than the parts of the map with a low economic potential.
Adapted from 'Industrial location and economic potential' by Colin Clark (1966), *Lloyd's Bank Review*, 82, 1–17

Figure 2.7b Motorways and dual carriageways in the United Kingdom, 1975
Reprinted with permission from *The UK Space* (2nd edition) edited by J. W. House (1977), published by Weidenfeld and Nicholson, London, figure 5.3

Figure 2.7c Railways in the United Kingdom, 1975
Reprinted with permission from *The UK Space* (2nd edition) edited by J. W. House (1977), published by Weidenfeld and Nicholson, London, figure 5.5

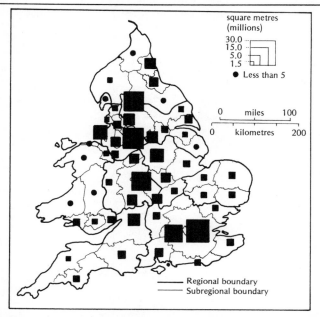

Figure 2.8   Subregional manufacturing floorspace in England and Wales, 1964
Source of data: Unpublished Department of Environment statistics
Reprinted with permission from *Industrial Location and Planning in the United Kingdom* by
D. E. Keeble (1976), published by Methuen & Co Ltd, London, figure 2.2a

in places declining, population. Average incomes are low, unemployment rates high, and many
people move away to economically healthier climes. Its component regions are too remote from
the core of purchasing power to attract new industries and branch plants of parent firms cannot
run efficiently in them. The periphery as a whole is relatively devoid of entrepreneurial talent and
fails to provide an environment conducive to new enterprise. The dearth of local entrepreneurs is
one reason for the slow pace of industrial development in Northern Ireland since the war. It is also,
compared with the core region, deficient in transport services (Figures 2.7b and c) and in its
supply of labour. D. H. Green has examined the movement of 82 manufacturing firms from the
South East and West Midlands to the South West, North East, and South Wales during the period
1967 to 1970. He asked firms to comment on their old and new locations.

1   Examine Figure 2.9 and comment on the differences in labour between old and new locations.

Despite the apparent deficiencies of peripheral regions, some studies have suggested that, in the
long run, most manufacturing industries could operate with success in them. The Toothill Committee
(1961) made a survey of the Scottish economy and found that high transport costs would be offset
by, among other things, lower costs of labour. And indeed a little expansion of manufacturing,
especially light manufacturing, has taken place in peripheral regions. Certainly some industries,
including the manufacture of electrical goods, seem to have made a go of it. The development of
the periphery has been brought about partly by a growing affluence in the United Kingdom as a
whole and partly by action taken by the government (which will be discussed a little later).

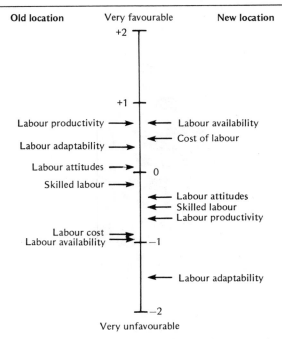

Figure 2.9   The evaluation of labour factors in old (central) and new (peripheral) regions. The scale measures the favourability of the factors mentioned.
Adapted from the work of D. H. Green (1974), 'Information, perception and decision making in the industrial relocation decision,' University of Reading, unpublished Ph.D. thesis

THE REGIONAL PROBLEM

The so-called 'regional problem' is that there is an imbalance of industrial and economic activity between core and periphery. The actual problem lies in the periphery where slow, stagnant, or declining industries have led to an unhealthy economic climate. Various indices can be used to take the 'economic temperature' of a region, including the percentage of unemployed workers and the dependency ratio (conventionally measured by the proportion of those aged 0–14 years and of retirement age, the nonactive population, as related to the 15–64 age group, the active population).

1   Table 2.1 lists some measures of **economic health** in the regions of the United Kingdom.

a   Make a choropleth map for each measure (Figure 2.10).

b   Discuss the extent to which the patterns you have mapped reflect differences between core regions and peripheral regions.

In trying to explain the causes of industrial and economic imbalance between regions in the United Kingdom, two schools of thought have grown up. The **structural school** argues that the economic problems of peripheral regions are due to their containing overspecialized, heavy industries now in decline, including steel, shipbuilding, textiles, and agriculture, which were inherited from the last century.

Table 2.1   Some measures of economic health in the United Kingdom, 1979

Region	Gross weekly earnings		Unemployment		Population change 1976–1977		Net migration 1976–1977		Dependency rate Nonactive/Active	Employment in agriculture	
	£	Rank	Percent	Rank	Percent	Rank	Thousands	Rank	Rank	Percent	Rank
South East	84.1	1	4.5	1	+0.3	3	−29.9	11	1.10 2	1.1	2
East Anglia	72.6	10	5.3	3	+0.7	1	+11.3	2	1.19 7	7.0	10
South West	73.4	9	6.9	6	+0.5	2	+23.5	1	1.28 9	4.6	7
East Midlands	74.5	8	5.1	2	+0.2	4	+5.3	3	1.20 8	6.8	9
West Midlands	76.3	6	5.8	4=	−0.2	7	−9.2	8	1.09 1	2.2	3
Yorkshire and Humberside	75.4	7	5.8	4=	−0.1	6	−1.5	5	1.16 4	2.5	4
North West	76.8	4	7.5	7	−0.3	9=	−14.4	10	1.15 3	0.9	1
North	77.5	3	8.4	10	−0.3	9=	−5.0	6	1.18 6	2.7	5
Scotland	78.3	2	8.3	9	−0.2	7	−9.8	9	1.17 5	4.5	6
Wales	76.5	5	8.1	8	0.0	5	+2.6	4	1.30 10	6.3	8
Northern Ireland	72.1	11	11.2	11	No data	—	−8.2	7	1.47 11	7.3	11

Source of data: Regional Statistics No. 14, H.M.S.O., London, 1979

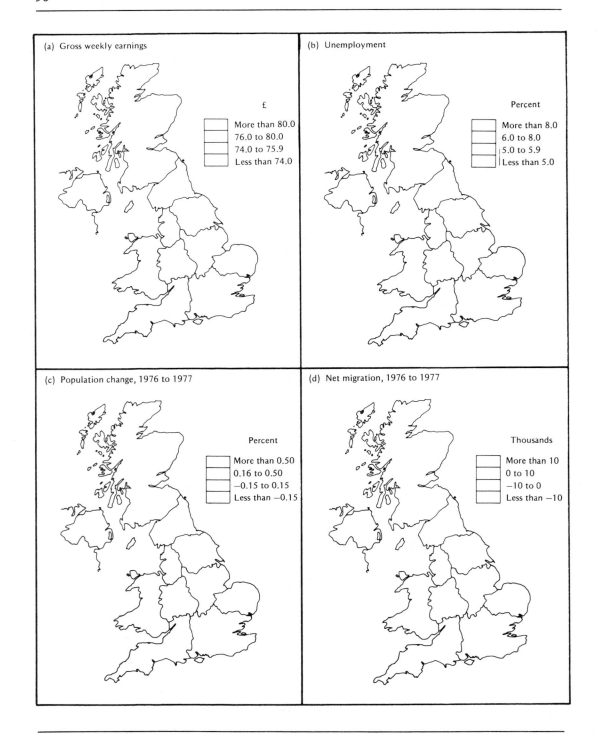

(a) Gross weekly earnings

£

More than 80.0
76.0 to 80.0
74.0 to 75.9
Less than 74.0

(b) Unemployment

Percent

More than 8.0
6.0 to 8.0
5.0 to 5.9
Less than 5.0

(c) Population change, 1976 to 1977

Percent

More than 0.50
0.16 to 0.50
−0.15 to 0.15
Less than −0.15

(d) Net migration, 1976 to 1977

Thousands

More than 10
0 to 10
−10 to 0
Less than −10

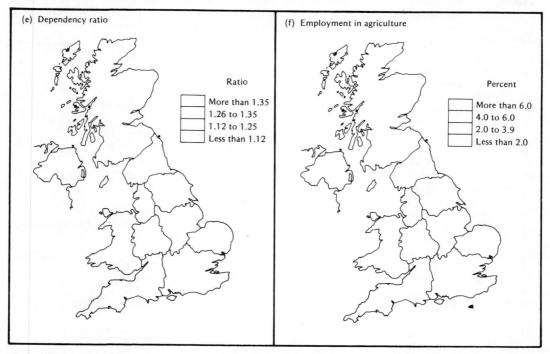

Figure 2.10   Some measures of economic health in the United Kingdom, 1979. (The regions are as on Figure 1.17.)

2 Table 2.2 shows industries in the South East and in the North placed in rank order on the basis of number of employees in employment in June 1976. Plot the data on Figure 2.11 and then compare and contrast the rankings in the two regions.

The problem with heavy industries is that, in times of **economic recession**, it is they, rather than consumer industries, which tend to feel the pinch first as demand falls. The result is that during recessions the unemployment rates in peripheral regions shoot up.

Table 2.2   Industries in the South East and the North of England ranked in order of employees in employment in June 1976 (highest is rank 1)

Rank	South East	North
1	Professional and scientific services	Professional and scientific services
2	Distributive trades	Distributive trades
3	Miscellaneous services	Miscellaneous services
4	Transport and communication	Construction
5	Public administration and defence	Public administration and defence
6	Insurance, banking, finance, and business services	Mechanical engineering
7	Construction	Transport and communication
8	Electrical engineering	Chemicals and allied industries
9	Paper, printing, and publishing	Mining and quarrying
10	Mechanical engineering	Shipbuilding and marine engineering
11	Vehicles	Metal manufacture
12	Food, drink, and tobacco	Electrical engineering
13	Chemicals and allied industries	Food, drink, and tobacco
14	Metal goods not elsewhere specified	Insurance, banking, finance, and business services
15	Gas, electricity, and water	Clothing and footwear
16	Other manufacturing industries	Textiles
17	Timber, furniture, etc.	Paper, printing, and publishing
18	Agriculture, forestry, fishing	Gas, electricity, and water
19	Clothing and footwear	Agriculture, forestry and fishing
20	Instrument engineering	Metal goods not elsewhere specified
21	Bricks, pottery, glass, cement, etc.	Other manufacture
22	Shipbuilding and marine engineering	Bricks, pottery, glass, cement, etc.
23	Metal manufacture	Timber, furniture, etc.
24	Textiles	Vehicles
25	Mining and quarrying	Instrument engineering
26	Coal and petroleum products	Coal and petroleum products
27	Leather, leather goods, and fur	Leather, leather goods, and fur

3a Using the data listed in Table 2.3, construct a scatter-graph (Figure 2.12) of unemployment for the period 1949 to 1979 by plotting unemployment rates for the North against national unemployment rates.

b Draw a line of best fit for the Northern region. (A line of best fit for data for the South East region has been plotted for you.)

c If the British economy were to fall into a recession and national unemployment rates were to increase, what would be the likely effect on unemployment rates in (i) the South East and (ii) the North? (One way to answer this question is to work out from the best-fit lines the percent increase in unemployment in the South East and in the North when the national unemployment rate increases by 1.0 percent.)

The second school of thought concerning the reasons behind the regional imbalance of industry is the **locational school**. Its adherents argue that the peripheral location of depressed regions is to blame. According to David Keeble, six chief factors, all of which vary from one region to another, fuel the core-periphery engine, lead to a regional imbalance of industry, and explain the locational change of manufacturing industry in the United Kingdom since the turn of the century. The six

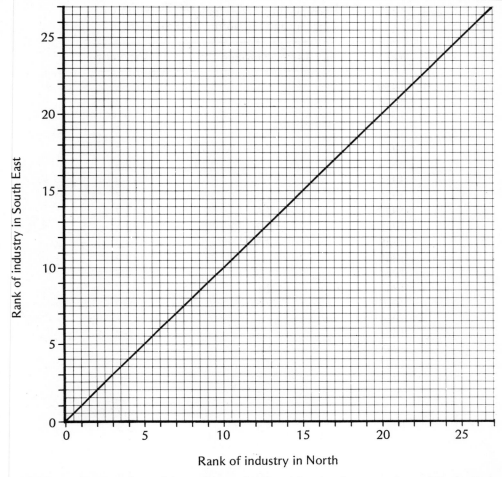

Figure 2.11 A comparison of the rankings of industrial orders in two regions of the United Kingdom. The 45-degree line indicates equal ranks in both regions. The further away from this line an industry is, the greater its rank importance either in the South East (if it lies above the line) or in the North (if it lies below the line).

factors are access to markets, leadership in developing new technology, the quality of transport links, agglomeration economies, the supply of labour, and entrepreneurial and institutional attitudes and skills.

If the structural school is right, then it is worth trying to attract industry to peripheral regions, despite their historical legacy of spoil heaps, slums, and derelict mills. On the other hand, if the locational school is right, then industrial expansion in peripheral regions is an economically unsound proposition: industries moving to the periphery face high costs of manufacture and marketing. Whichever view is nearer the truth, attempts have been made by successive governments to take industry from the industrial core to peripheral areas.

Table 2.3   Unemployment rates (percent) in the United
Kingdom and the North, 1949 to 1979

Year	United Kingdom	The North
1949	1.6	2.6
1950	1.6	2.8
1951	1.3	2.2
1952	1.2	2.6
1953	1.8	2.4
1954	1.5	2.3
1955	1.2	1.8
1956	1.3	1.5
1957	1.6	1.7
1958	2.2	2.4
1959	2.3	3.3
1960	1.7	2.9
1961	1.6	2.5
1962	2.1	3.7
1963	2.6	5.0
1964	1.7	3.3
1965	1.5	2.6
1966	1.6	2.6
1967	2.5	4.0
1968	2.5	4.7
1969	2.5	4.8
1970	2.6	4.7
1971	3.5	5.8
1972	3.9	6.4
1973	2.7	4.7
1974	2.6	4.6
1975	4.1	5.9
1976	5.7	7.4
1977	6.2	8.3
1978	6.1	8.8
1979	5.7	8.5

Source of data: *British Labour Statistics — Historical
Abstract, 1886–1973*, H.M.S.O., London, 1974–1979,
Department of Employment, Statistics Division C1,
Watford

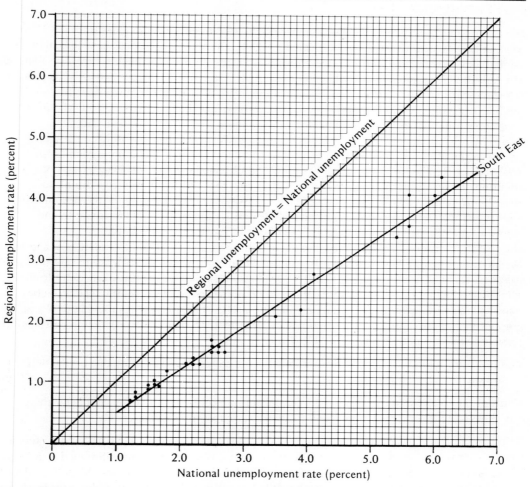

Figure 2.12   Unemployment rates in the United Kingdom and two of its regions, 1949 to 1979

GOVERNMENT ACTION

Direct government response to regional imbalance in the United Kingdom dates from the 1930s when unemployment reached unprecedentedly high levels. Areas which specialized in coal-mining, shipbuilding, textiles, and iron and steel experienced rates of unemployment in excess of 30 percent. The Industrial Transference Board had been set up in 1928 to retrain labour and assist its transfer to regions where it was needed. By 1938, when the Board was abolished, it had been responsible for 200,000 transfers. But it was the Special Areas Act of 1934 which marked the beginning of a regional policy which aimed at taking the work to the workers. Between then and 1978, government financial commitment increased substantially, as did the areas which qualify for assistance (Figure 2.13). By 1978, 43 percent of Britain's working population lived in assisted areas. The present government, which has decided on a regional policy involving a focus of attention on much smaller areas of development, hopes that this figure will have shrunk to a mere 25 per cent by 1983. A summary of regional assistance offered by the British government is given in Table 2.4a. The measures listed include positive inducements, such as grants, and negative controls, such as the Industrial Development Certificate (Table 2.4b).

The operation of the measures can be illustrated with the aid of D. M. Smith's space-cost and space-revenue curves (p. 65). Total costs consist of production costs, which we shall assume are the same everywhere, and distribution costs, which we shall assume increase at a steady rate away from the point of production (point A) (Figure 2.14a). Under these conditions, a central region can be identified in which manufacturers can operate with profit. If a government should impose a tax on firms located in the centre of the central region, the effect will be to raise production costs and so possibly to make production unprofitable (Figure 2.14b).

1a Modify Figure 2.14a to show how a government could create a zone of profitable production around site B.

 b Which of the two measures — making the central area unprofitable or making peripheral areas profitable — is a government more likely to adopt? Give reasons for your choice.

Despite more than forty years of regional policy in the United Kingdom, industrial and economic inequality between regions remains. One of the worst problems has been the continued high levels of unemployment in some areas. But the fact that high unemployment rates are still with us does not necessarily mean that regional policy has failed. It can be argued that unemployment levels in problem regions would have been much higher in the absence of a regional policy. Furthermore, much movement of the industrial workforce has occurred in response to government measures. During the period from 1966 to 1971, an average of 200 manufacturing firms were moving each year to different regions and London and the West Midlands actually recorded a decline in manufacturing employment. It is of interest that the problem regions given the biggest 'handouts' attracted more industry than others. Part of the success of government action can be seen in the relocation of firms.

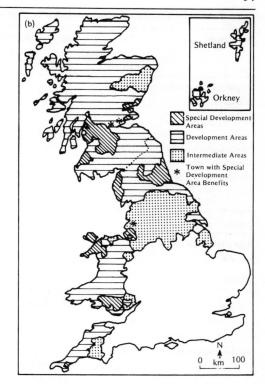

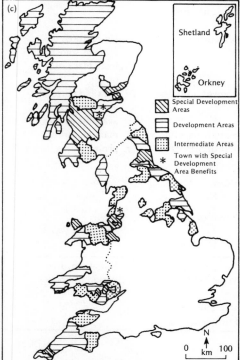

Figure 2.13   Government-assisted areas
(a) The Special Areas of 1934 to 1939
(b) Assisted Areas as defined by the
Department of Industry to take effect in 1979
(c) Assisted Areas as defined by the Department
of Industry to take effect in 1982

Table 2.4a   Summary of government incentives for industry in Areas of Expansion

Incentive	Special Development Areas (SDAs)	Development Areas (DAs)	Intermediate Areas (IAs)	Northern Ireland
Regional development grants:				Capital and industrial development grants
New buildings and works (other than mining works) and adaptations	22%	15%	Nil	
New machinery and plant	22%	15%	Nil	30% to 50%
Selective financial assistance				
Loans	On favourable terms for general capital purposes for projects which provide additional employment or safeguard significant existing employment.			On favourable terms for general purposes. Loan guarantees may also be negotiated.*
Interest relief grants*	As an alternative to loans on favourable terms, grants towards the interest costs of finance provided from nonpublic sources for projects which provide additional employment, or safeguard significant existing employment.			As in SDAs, DAs and IAs.
Removal grants*	Grants of up to 80% of certain costs incurred in moving an undertaking into one of these Areas.			Up to 100% of costs of transfer from place of origin.
Service industry grants*	To help offices, research and development units and other service industry undertakings. A grant of £1,500 for each employee moving with his work into the Areas (within a limit of 50% of new jobs created).			Flexible range of assistance available as for manufacturing projects.
	A grant for each job created in the Areas: £1,500	£1,000	Nil	
	Grant to cover the cost of approved rent of premises in the new location for: Up to 7 years	Up to 5 years	Up to 3 years	
Government factories	For rent: standard factories. Rent-free period, may be up to 5 years in Special Development Areas and up to 2 years elsewhere.* 99 years' lease of standard and custom-built factories may be purchased. Payment spread over a period.			Initial rent-free period of up to 5 years.*

Incentive	Special Development Areas (SDAs)	Development Areas (DAs)	Intermediate Areas (IAs)	Northern Ireland
Employment grants	Nil	Nil	Nil	A substantial contribution towards setting-up costs of attractive projects.
Selective employment payment to manufacturers	Nil	Nil	Nil	£2 weekly for full-time adult employees.
Training services	Free training services operated by the Training Services Agency.			Free courses at Government Training Centres; grants for training on employers' premises.
Help for transferred workers	Settling-in grants, separation and disturbance allowances, and substantial help with travel and removal expenses.			Full fares and household removal costs or lodging allowances plus substantial settling-in grants for key workers.
Taxation	Regional development grants in Great Britain and the corresponding grants up to 45% in Northern Ireland are not treated as reducing capital expenditure in computing tax allowances. Other grants are treated as trading receipts.			
Finance from European Community funds	Loans may be available on favourable terms from the European Investment Bank and the European Coal and Steel Community. Loans for employment-creating projects in the Special Development Areas, Development Areas and Northern Ireland, may be guaranteed by the government against the foreign exchange risk. Under a special agency arrangement EIB loans from £30,000 to £2.6m can be arranged through the appropriate government department.			
Contracts preference schemes	Preferential treatment when tendering for contracts placed by government departments, nationalized industries, etc.		Nil	As in SDAs and DAs.

*Incentives marked with an asterisk are subject to the provision of sufficient additional employment to justify the assistance sought.

Table 2.4b Industrial development certificates

When required	Application for planning permission for industrial development must usually be supported by an industrial development certificate (IDC). IDCs are, however, **not** required in **Development Areas, Special Development Areas,** or **Northern Ireland.**
	Outside these Areas an IDC is needed if the industrial floor space to be created by the development (together with any related development) exceeds 12,500 square feet (1,162 square metres) in Southeastern England, and 15,000 square feet (1,395 square metres) elsewhere. IDC applications should be made to the offices of the government department responsible for the area in which the project is proposed.
Intermediate Areas	IDCs are generally freely available in the Intermediate Areas for the expansion of long-established firms. Newcomers may have to show that they could not reasonably be expected to go to a Special Development Area or a Development Area; the Department may also wish to discuss in which particular location the applicant's needs for labour can best be met.
Outside the Areas for Expansion	Applications for locations outside the Areas for Expansion are examined more critically with special reference to the extent to which the development is considered to be mobile as to location and is compatible with the needs and resources (particularly labour) of the locality concerned. Where a project is considered to be mobile, the IDC may be refused and the applicant encouraged to carry out the project in one of the Areas for Expansion. Certain inner areas of London and Birmingham, followed by the active new and expanded towns, take precedence after the Areas for Expansion in consideration of IDC applications for mobile projects coming forward from the relevant region.

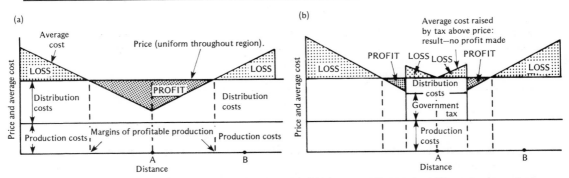

Figure 2.14 The effect of a government tax on spatial margins of profitability

THE RELOCATION OF MANUFACTURING FIRMS

Movement in manufacturing industry from the South East region to peripheral areas during the period 1945 to 1965 is shown in Figure 2.15. Examining the map closely, it can be seen that the **friction of distance** seems to influence the amount of movement: more firms move to closer parts of the periphery than to the remoter parts. Presumably, too, the attractiveness of a peripheral region, as measured by the availability of labour in it (that is, the number of unemployed), will influence the amount of **relocation**. This hypothesis can be put to the test by applying a simple **gravity model** (see *Settlements*, p. 94) written as:

$$M_{ij} = G(A_j/d_{ij})$$

where M_{ij} is a measure of the predicted volume of industrial movement between region i, the source area from which the industry is moving, and region j, the destination of the industry which is moving; G is a constant which will be defined a little later; A_j is the availability of labour (the number of unemployed) in region j, weighted to allow for changes in unemployment during the twenty-year period of study; and d_{ij} is the shortest road distance between the centre of region i and the town nearest the industrial centre of gravity of region j.

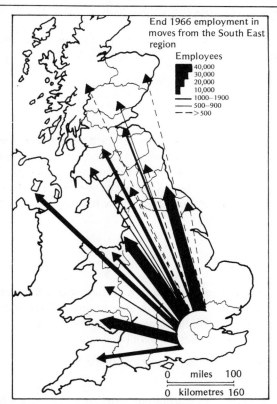

Figure 2.15   Manufacturing movement to peripheral areas from the South East region, 1945 to 1965

Reprinted with permission from 'Industrial movement in the United Kingdom' by D. E. Keeble (1972), *Town Planning Review*, 43, 3–25, published by Liverpool University Press, figure 8

A worked example: predicted movement of jobs from the South East region to all peripheral regions. (Follow the steps through with Table 2.5a)

Step one For each peripheral region (numbers 2 to 19, column 1), calculate the predicted movement index, I_j (column 4) by dividing the weighted unemployed workforce index, A_j (column 2), by the road distance, d_{ij} (column 3). For Northern Ireland, denoted by the subscript 2, and with the South East denoted by the subscript 1, we have

$I_2 = A_2/d_{12}$
$\quad = 82{,}139/1056$
$\quad = 78$

Step two Add up the predicted movement indices for all peripheral regions. In other words, find the total of column 4, $\Sigma_{j=2}^{19} I_j$. (As was mentioned earlier, the Greek captial sigma, Σ, simply means 'sum of'. The numbers below and above it simply indicate that the values in regions 2 to 19 are to be summed.) The answer in the example is 1104.

Step three Compute the total number of jobs actually created in all peripheral regions, $\Sigma_{j=2}^{19} J_j$, by finding the total of column 6. The answer in the example is 168.4 (thousand).

Step four Compute the constant G by dividing the total of the predicted movement indices, $\Sigma_{j=2}^{19} I_j$, by the total number of jobs actually created, $\Sigma_{j=2}^{19} J_j$:

$G = (\Sigma_{j=2}^{19} I_j)/(\Sigma_{j=2}^{19} J_j)$
$\quad = 168.4/1104$
$\quad = 0.1525.$

Step five Calculate the predicted movement of jobs in each peripheral region, M_{ij} (column 5) in the following way. For each region, take the predicted movement index, I_j (column 4) and multiply by G. This gives the predicted movement of jobs, in thousands, from region i (in the example $i=1$, the South East) to region j (one of the 18 peripheral regions). Put the value in column 5. In the case of predicted movement to Northern Ireland, we have

$M_{12} = G \times I_2$
$\quad = 0.1525 \times 78$
$\quad = 11.9 \text{ (thousands)}.$

Figure 2.16a shows the predicted movement of jobs plotted against the number of jobs actually created. If predictions made a perfect match with observations, all points would lie on the 45-degree line. That the points lie near this line indicates that the simple gravity model predicts well the movement of jobs from the South East to peripheral regions.

2a Apply the simple gravity model to the data given in Table 2.5b which is for movement from the West Midlands to peripheral regions.

b Plot your results on Figure 2.16b. (*Note*: values of zero cannot be plotted on logarithmic coordinates.)

c How good is the model as a predictor of movement of jobs from the West Midlands?

Table 2.5a Gravity model data and movement from the South East (region 1) to peripheral regions, 1945 to 1965

j Peripheral Region	2 Weighted unemployed workforce index, A_j	3 Road distance, d_{ij} (km)	4 Predicted movement index, $I_j = A_j/d_{ij}$	5 Predicted movement, M_{ij}† (thousands)	6 Actual number of jobs created, J_j (thousands)
2 Northern Ireland	82,139	1056*	78	11.9	16.9
3 Central Scotland	95,697	634	151	23.0	13.4
4 South West Scotland	5,296	596	9	1.4	1.9
5 Eastern Borders	526	552	1	0.2	0.0
6 East-central Scotland	21,907	633	35	5.3	7.5
7 East Scotland	8,824	686	13	2.0	0.9
8 North East Scotland	9,624	797	12	1.8	0.4
9 Highlands and Islands	7,647	855	9	1.4	0.4
10 South Wales	65,896	264	250	38.1	36.3
11 North East Wales	6,637	288	23	3.5	5.6
12 Rural Wales	9,559	298	32	4.9	1.3
13 North East England	71,354	417	171	26.1	34.4
14 West Cumberland	3,493	481	7	1.1	0.9
15 North Yorkshire	2,134	362	6	0.9	0.2
16 Northumbria	1,524	489	3	0.5	0.1
17 Rural Cumberland and Westmoreland	1,650	449	4	0.6	0.2
18 Merseyside and Southwest Lancs	79,758	317	252	38.4	36.0
19 Devon and Cornwall	16,276	340	48	7.3	12.0
Totals			$\sum_{j=2}^{19} I_j = 1104$	$\sum_{j=2}^{19} M_{ij} = 168.4$	$\sum_{j=2}^{19} J_j = 168.4$

*This has been arbitrarily lengthened to allow for transport problems involved in crossing the Irish Sea.

†This is defined by the equation $M_{ij} = G \times I_j$.

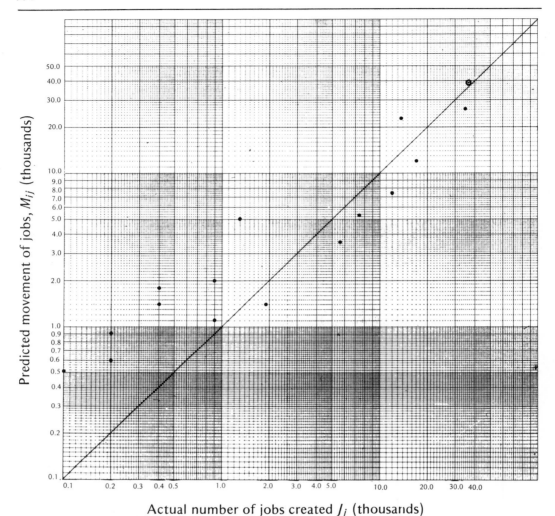

Figure 2.16a Movement of jobs, M_{ij}, predicted by a simple gravity model for movement from the South East region to peripheral regions (Table 2.5a), plotted against the actual number of jobs created in peripheral regions from the South East. If predictions were 'spot on', all points would lie on the 45-degree line.

3 Draw up your results of predicted movement of jobs from the West Midlands as a flow diagram (Figure 2.17). (See Figure 2.15 for an example.)

4 What conclusions can you draw about the effect of labour availability (as measured by unemployment) and distance on the movement of jobs from the core regions of the United Kingdom to peripheral regions?

5 Why might a firm be loath to move from a core region? What advantages might there be in staying put?

Table 2.5b Gravity model data and movement from the West Midlands region (region 1) to peripheral regions, 1945 to 1965

	1 Peripheral region	2 Weighted unemployed workforce index, A_j	3 Road distance, d_{ij} (km)	4 Predicted movement Index, $I_j = A_j/d_{ij}$	5 Predicted movement, M_{ij} (thousands)	6 Actual number of jobs created, J_j (thousands)
2	Northern Ireland	82,139	808			5.7
3	Central Scotland	95,697	462			7.4
4	South West Scotland	5,296	440			0.3
5	Eastern Borders	526	414			0.0
6	East-central Scotland	21,907	494			0.6
7	East Scotland	8,824	551			0.3
8	North East Scotland	9,624	658			0.1
9	Highlands and Islands	7,647	716			0.0
10	South Wales	65,896	155			15.3
11	North East Wales	6,637	116			0.9
12	Rural Wales	9,559	147			1.6
13	North East England	71,354	298			4.8
14	West Cumberland	3,493	311			0.0
15	North Yorkshire	2,134	254			0.2
16	Northumbria	1,524	370			0.0
17	Rural Cumberland and Westmoreland	1,650	279			0.0
18	Merseyside and Southwest Lancs.	79,758	153			39.0
19	Devon and Cornwall	16,276	327			1.3
				$\sum_{j=2}^{19} I_j =$	$\sum_{j=2}^{19} M_{ij} =$	$\sum_{j=2}^{19} J_j =$

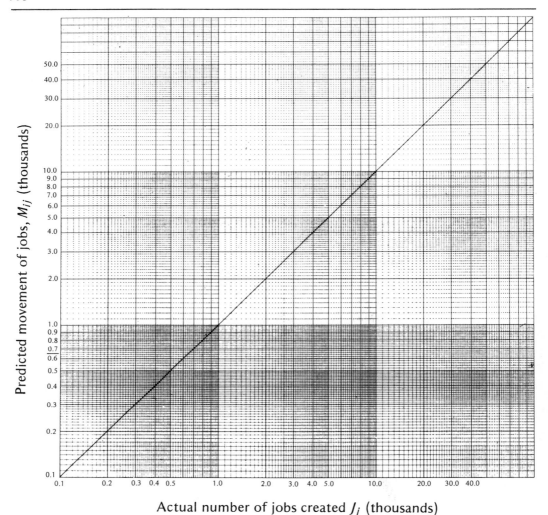

Figure 2.16b Movement of jobs, M_{ij}, predicted by a simple gravity model for movement from the West Midlands region to peripheral regions (Table 2.5b), plotted against the actual number of jobs created in peripheral regions from the West Midlands. If predictions were 'spot on', all points would lie on the 45-degree line.

Figure 2.17   Flow map of the predicted movement of jobs from the West Midlands to peripheral regions. The regions are as follows (cf. Table 2.5b): 1 West Midlands; 2 Northern Ireland; 3 Central Scotland; 4 South West Scotland; 5 Eastern Borders; 6 East-Central Scotland; 7 Eastern Scotland; 8 North East Scotland; 9 Highlands and Islands; 10 South Wales; 11 North East Wales; 12 Rural Wales; 13 North East England; 14 West Cumberland; 15 North Yorkshire; 16 Northumbria; 17 Rural Cumberland and Westmoreland; 18 Merseyside and South West Lancashire; 19 Devon and Cornwall

NORTHERN ENGLAND: A CASE STUDY IN INDUSTRIAL DEVELOPMENT

The northern region of England, which comprises the counties of Cumbria, Northumberland, Durham, Cleveland, and the metropolitan county of Tyne and Wear, contains 5.6 percent of the United Kingdom's 56 million inhabitants.

The region prospered during the nineteenth century when coal-mining, iron and steel, shipbuilding, and engineering flourished. During the 1920s, heavy industry developed on Teesside but by the early 1930s the prosperity of the region had waned. In 1934, parts of the region were bad enough to be designated Special Areas by the government (Figure 2.13a). But since then, many of the basic industries in the region have continued to decline. The coal-mining industry, formerly the largest employer in the area, now employs a mere 34,000 workers in the surviving collieries near the coast. The iron and steel industry has, by and large, moved to the coast; the recent completion of the Redcar scheme threatens the future of the Consett works. Shipbuilding, despite the restructuring initiated by the Geddes Report in 1966 and financial assistance from the government, has failed to cope with fierce foreign competition and a shrinking world market; demand for offshore oil platforms has turned out to be short-lived. One industry which has expanded is the chemical industry. Initially based on the production of alkali at Tyneside, the heavy chemical industry and, more recently, the petrochemical industry are now concentrated at Teesside where investment continues. Imperial Chemical Industries have recently completed a £150 million ethylene plant at Wilton.

The northern region has undoubtedly benefited from regional aid — the growth of the Gross Domestic Product per capita was a little above the national average in the early 1970s. However, the region has tended to attract industries which need a lot of capital but not much in the way of labour. Fairly high levels of employment were generated while the industrial plants were being built, but after completion relatively few permanent jobs were created. Efforts to boost the regional economy have been made at a local level. Tyne and Wear county, in an attempt to attract more light industry, has used its rates funds to build small factories and give grants and loans to firms. There has also been a major effort to encourage office employees to move to the region. Some 3000 government clerical jobs were created in the region between 1963 and 1970. Barclaycard have recently set up a regional centre in Cleveland.

Industrial developments have not been confined to long-established towns. The new towns of Peterlee, Aycliffe, and Washington have succeeded in attracting new industry on the strength of generous regional assistance and modern homes and factories. The completion of the M6 and A1(M) roads, as well as inter-city rail services, freightliner services, and improved port facilities at Teesport have given the region better accessibility to the national market (Figure 2.18).
So the economic outlook for the North is, on the face of it, not too bleak. But, despite these developments, unemployment remains above the national average and people continue to move away, especially from the rural parts where old folk make up an unusually large proportion of the population. The region has never made a big contribution to the national output of agricultural produce, the upland terrain, generally poor soils, and inclement climate making farming difficult.

Farming on the hills has always been a touch-and-go business and relies heavily on subsidies to keep it going. The area under forest has been increased greatly in recent years. As well as providing timber, the forests have played a part in the growth of a tourist industry. Forest parks have been established along with national parks and areas of outstanding natural beauty. The tourist industry has also been aided by the 1969 Development of Tourism Act which provides funds for the building of hotels in the region.

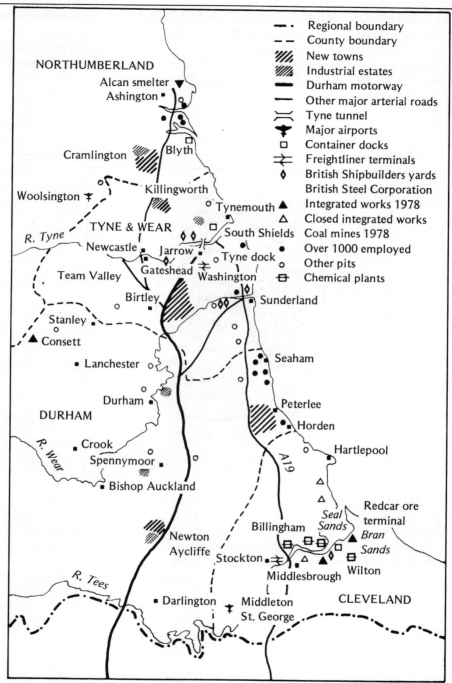

Figure 2.18   North East England: the economic core
Reprinted with permission from *Regional Development in Britain* (second edition) by
G. Manners *et al.*, published by John Wiley and Sons, Chichester, 1980, figure 13.2.
Copyright © 1980 John Wiley & Sons.

Northern England not only lies peripheral to the industrial core of the United Kingdom, it also lies peripheral to the industrial core of Europe (Figure 2.19). Industry in all peripheral parts of Europe faces problems. Problem areas within the European Community are offered financial assistance, over and above that offered by member countries, from the European Regional Development Fund and the European Investment Bank (Figure 2.20).

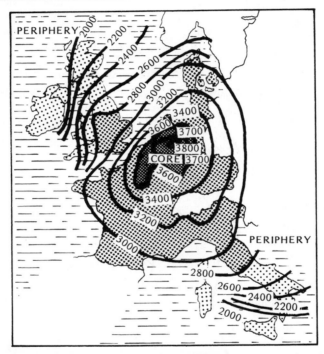

Figure 2.19   A map of economic potential in the European Community. It reveals a core-periphery pattern. The core region, called the 'Golden Triangle', consists of the Rhine-Ruhr region, northeastern France, and parts of the Benelux countries. It is surrounded by an extensive periphery which includes northern England and southern Italy.

Though industrial growth is hampered by problems in all peripheral regions of Europe, the set of problems it faces varies from one part of the periphery to another. We have described the problems of industrial growth in northern England and we shall now turn our attention to industrial development in southern Italy, another peripheral region of Europe.

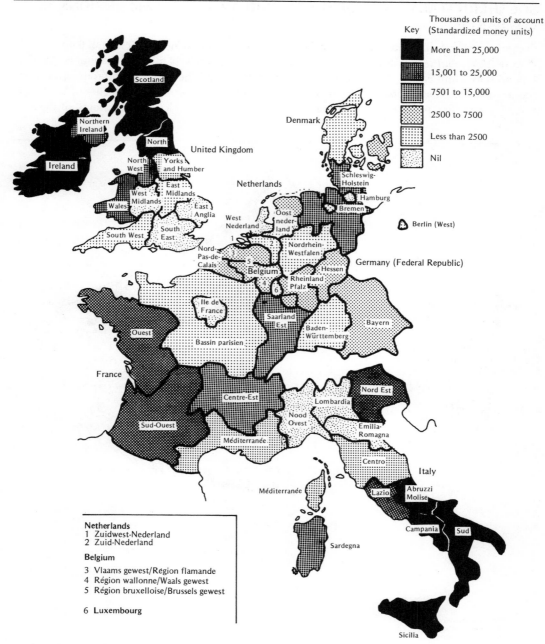

Figure 2.20   Investment grants in regions in the European Community

SOUTHERN ITALY AND NORTHERN ENGLAND: PROBLEM REGIONS COMPARED

Comprising all the mainland south of the Centro region and the islands of Sicily and Sardinia (Figure 2.21), southern Italy contains 37 percent of the country's 54 million inhabitants. At the time of Italy's unification, 1861, the economic development of the northern and southern regions was already unbalanced. Southern Italy had been for centuries exploited by the élite of the feudal system. Even when the feudal system was abolished in 1866, the agricultural economy of the area

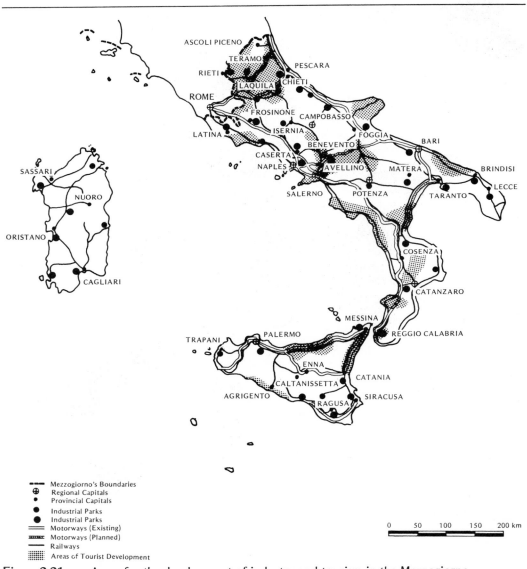

Figure 2.21 Areas for the development of industry and tourism in the Mezzogiorno

did not improve; a rural middle class emerged which merely perpetuated the inefficient farming of large estates (*latifundia*). Large-scale industrialization came to Italy towards the end of the nineteenth century. It took place largely in the northern region. This was because the northern parts possessed several advantages: a long cultural tradition, a skilled workforce, reserves of energy resources (oil, gas, and potential for hydro-electric power), a relatively efficient network of roads and railways, and ready access to European markets. The northern Italian core region grew at the expense of the southern Italian peripheral region, backwash effects creating a widening gap in the economic development of the two areas.

Effective attempts to redress the imbalance between the North and the South were not forthcoming until 1950. Improvements in the agricultural sector of the southern Italian economy were initiated by land reforms (see *Agriculture*, p.69). The large estates were expropriated and over 500,000 hectares of land were redistributed to 91,000 families of peasants. The *Cassa per il Mezzogiorno*, a central planning body which made funds available for land improvement schemes, afforestation, and the building of sewers and roads, was set up in the same year.

The physical geography of southern Italy hindered agricultural development. It is a mountainous area with generally poor soils. Very dry summers severely restrict the range of crops which may be grown. On the other hand, the guarantee of hot, dry summer days has been an important factor in the growth of a tourist industry. Since 1965, loans have been available for hotel construction. Ski resorts have been set up in the mountains of Calabria.

In 1957, the *Cassa per il Mezzogiorno* turned its attention to industrial development. It introduced cheap loans, tax concessions, and grants for plants and machinery. These measures were later reinforced by the Italian government's compelling a number of state industries, including iron and steel, to devote 60 percent of their industrial investment to the South (Table 2.6). Surprisingly, there are a number of industries, including cement-making, stone-quarrying, oil-refining, the production of chemicals, paper-making, and nonferrous metallurgy, which are not eligible for 'soft loans' or cash grants except when prevailing market conditions necessitate importing the products of these industries.

More recently, the 1976 to 1980 Five-Year Plan, as well as providing direct financial incentives for industry, also involved spending much money on industrial back-up services including industrial parks and motorways. The plan also entailed a number of 'special development projects' which have included developing water resources, expanding the area of irrigated land, and promoting the production and marketing of livestock, forestry products, and citrus fruit. Considerable sums of money have also been spent on improving social and economic conditions in the metropolitan areas of Naples and Palermo. Despite these filips to industrial development, the region has failed to attract industries requiring large workforces. Between 1951 and 1971, the growth of industrial employment was smaller than in the northern region, its share of industrial employment actually falling from 16.4 percent to 14.6 percent. With the spectre of unemployment never far away, people are still moving to the more attractive northern region.

1   Compare and contrast the regional problems and regional policies of northern England and southern Italy.

Table 2.6 Summary of government incentives for industry in the Mezzogiorno

FINANCIAL INCENTIVES

Soft loans. Loans at a low interest rate are extended for industrial projects regardless of the amount of the money allotted for fixed investments; these loans however are limited to the first 30 billion lire of investments. Such loans are equal to 40% of fixed investment and inventory (raw or pretreated materials for an amount not exceeding 40% of fixed investment). The rate of interest is 30% of the official one, which is fixed periodically by the government.

The maturity of loans cannot exceed 15 years for new projects and 10 years for enlargement and modernization of existing facilities.

Cash grants. Cash grants are extended on the basis of the cost of buildings, connections to utilities, machinery and equipment (fixed investments) as indicated below:

Investment Size	Cash grant
— for fixed investments up to 2 billion lire	40%
— for the part of fixed investments exceeding 2 and up to 7 billion lire	30%
— for the part of fixed investments exceeding 7 and up to 15 billion lire	20%
For projects involving fixed investments over 15 billion lire	20%

The above-mentioned measures of cash grants can be raised by one-fifth when the investment project concerns priority sectors, according to government directives.

A further one-fifth increase can be granted for projects locating in particularly depressed areas. At any rate total financial incentives (soft loans and cash grants) are not to exceed 70% of fixed investment. The only admitted exceptions concern the above-mentioned cases of priority sectors and special locations for which, therefore, a total maximum of 86% is allowed.

FISCAL INCENTIVES

The following fiscal exemptions are granted to favour industrial investments in the Mezzogiorno:

— Ten-year exemption from local tax on company profits (ILOR - present average rate about 15%);

— Exemption from local tax on 70% of company declared profits reinvested in industrial projects located in the Mezzogiorno;

— Ten-year 50% exemption from national tax on company profits (IRPEG - present average rate about 25%).

INCENTIVES FOR RESEARCH AND DEVELOPMENT ACTIVITIES

Incentives are available, always in the frame of the Mezzogiorno development policy, to businessmen setting up R&D centres or undertaking R&D projects. For R&D centres, cash grants up to 50% of the fixed investment cost are available, provided at least 25 researchers are employed. R&D centres benefit from the reduction of social charges on salaries as mentioned above.

Further financial incentives are available for R&D activities according to the law which created the 'Fund for Research' managed by IMI (Instituto Mobiliare Italiano). Forty percent of this fund available assets is in fact to be assigned to projects carried on in the Mezzogiorno.

IMI is entitled:

— to take shares in enterprises engaged in research activities;

— to extend 8 to 10-year soft loans for R&D projects;

— to extend cash grants amounting to 40% of the research project costs.

INDUSTRIAL PARKS

170 industrial parks were created in the Mezzogiorno; they are provided with the services and the infrastructures necessary to allow the settlement of manufacturing enterprises under best conditions. Detailed information about such industrial parks can be obtained from IASM.

REDUCTION OF SOCIAL CHARGES

Industrial employers of the Mezzogiorno operating in the sectors indicated by the government are totally exempted from the payment of yearly contributions to the pension fund of workers for ten years to begin with the date of hiring of the same workers. Such payments in fact will be temporarily made by the government itself. This exemption involves an average reduction of about 18% in the overall labour cost.

Industry in Socialist States

In socialist countries, **state ownership** of industry is a powerful vehicle of industrial location. F. E. Iain Hamilton recognized four principles of industrial location followed by many socialist states. The first principle is that industries should be located near the sources of raw materials and fuel they use. The second is that industries should be located near the markets for their products. And the third is that industrial plants should be located so as to achieve an even spread of industry. This means that backward areas are developed so that, in the long run, all persons in the country receive the same income. It also means that differences in income levels between urban areas and rural areas and between ethnic groups are smoothed out. The fourth principle is that industry should be dispersed in the interests of security and national defence. The siting of iron and steel plants in Yugoslavia is an example of this principle in action (Figure 2.22).

To a certain extent, some of these principles contradict one another. Furthermore, the gap between the ideals of the state planner and the actual distribution of industry may be wide. But one pattern these principles do promote is a wide spread of industry with, especially in countries belonging to the Council of Mutual Economic Assistance (COMECON), each region tending to specialize in the making of a particular group of products. This is certainly true in eastern Europe where, as we shall see for the cases of Poland and the German Democratic Republic, major changes in the distribution of industry in pursuit of ideological principles have been witnessed in recent decades.

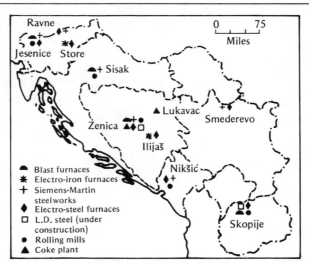

Figure 2.22   The location of plants in Yugoslavia built between 1945 and 1963. Yugoslavia's break with the Soviet Union in 1948 resulted in the cessation of financial aid, materials, and equipment. The iron and steel industry then had to watch costs very carefully. It also located new plants at Sisak, Ilijaš, Nikšić, and Skoplje, well away from the Soviet border.
Reprinted with permission from *Understanding Society* (1970), published by Macmillan for the Open University Press, essay by F. E. Ian Hamilton, figure 20.7

INDUSTRY IN POLAND

A. Kuklinski's study of Polish economic development highlights the sort of conflicting policies that emerged in socialist states after 1945. The Polish government followed policies which were to take industry from traditional centres and create employment in the underpopulated agricultural regions (Figures 2.23a and b). At the same time, the desire for rapid economic growth favoured investment in heavy industry in Upper Silesia, Poland's industrial heart. Moreover, a price system which kept the revenue from heavy industrial goods artificially low and other manufactured goods artificially high meant that heavy industry was forced to locate where costs were lowest. Membership of COMECON, an organization which fostered national specialization, also resulted in considerable investment in the traditional industrial centres, Poland specializing in the production of coal, chemicals, and primary metals. Since 1956, the emphasis in planning has shifted from ideology to

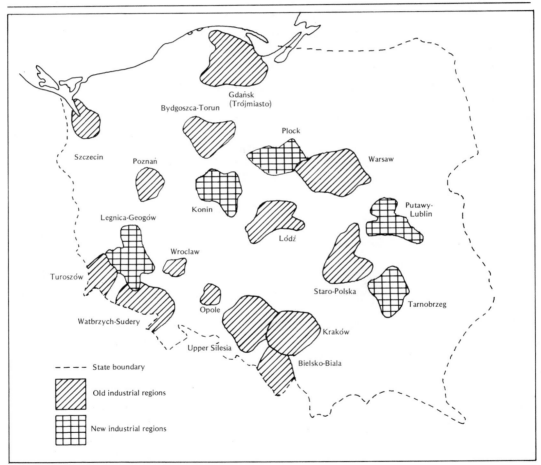

Figure 2.23a Industrial regions in Poland, 'old' and 'new'.
Reprinted with permission from 'Spatial industrial changes in Poland since 1945' by
R. Wikczewski, T. Lijewski, and B. Kortus (1978) in *Industrial Change: International Experience and Public Policy*, edited by F. E. Ian Hamilton, published by Longman, London, figure 8.2

economics. The desire to redress the industrial and economic imbalance within Poland remains an important element and major new industrial centres have been built where raw materials are found. The Konin region, for example, specializes in the production of power from large reserves of lignite (Figures 2.23a and b). Growth in old industrial areas has also occurred. A striking example is the growth of the iron and steel industry on the Upper Silesian coalfield since 1950 (Figure 2.23b). The Silesian iron and steel plants use local coking coal but import iron ore from the Soviet Union and from Sweden, though the works at Czestochowa, which were expanded in 1952, use local Jurassic ores. The new plant at Nowa Huta today produces over three times the Polish national output of 1938.

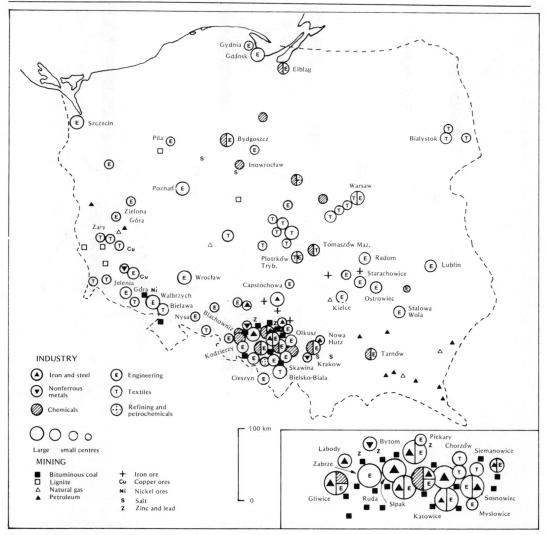

Figure 2.23b   The distribution of industry in Poland
Reprinted with permission from *Eastern Europe* by Roy E. H. Mellor (1975), published by Macmillan, London and Basingstoke, figure 9.3

1   Figure 2.24 shows the percent change in the proportion of people employed in industry by *voivedship* (region). Decide whether or not the map suggests that measures to decentralize industry in Poland have been successful. Briefly explain your decision.

We shall now test the hypothesis that, because most of Poland's raw materials and fuel supplies are located in Upper Silesia, and because of economic policies pursued since 1945, structural change of industry within the Upper Silesian region has been slow. We shall make the test firstly by constructing Lorenz curves and secondly by computing Conkling's index of industrial diversification.

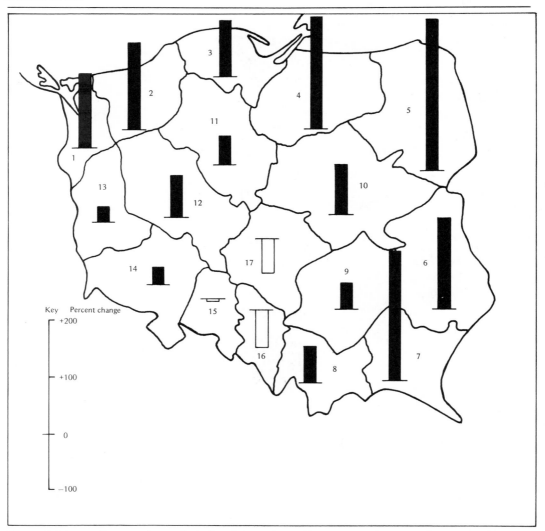

Figure 2.24   Percent change in industrial employment between 1951 and 1972. Regions: 1 Szczecin; 2 Koszalin; 3 Gdansk; 4 Olsztyn; 5 Bialystock; 6 Lublin; 7 Rzeszów; 8 Kraków; 9 Kieke; 10 Warszawa; 11 Bydgoszcz; 12 Poznan; 13 Zielona Góra; 14 Wrocław; 15 Opole; 16 Katowice; 17 Łódź.

Lorenz curves are used to gauge the mix of industries in a region. They enable the actual **industrial employment structure** of a region to be compared with the employment structure showing the greatest possible mix, that is, each industry taking the same share of the workforce in the region. The procedure, illustrated here by the data for industrial employment in Upper Silesia in 1946 (Table 2.7), runs like this:

i Employment figures for each industrial activity (16 groups are used in the example), expressed

Table 2.7 Employment in some extractive and manufacturing industries in Upper Silesia, Poland, 1946 and 1970

	Industry (in rank order)	Employment in each industry expressed as a percentage of the total employment in all industries	Cumulative percentage
1946	Mining	49.7	49.7
	Ferrous metallurgy	18.7	68.4
	Machinery and electrical engineering	11.4	79.8
	Nonferrous metallurgy	3.8	83.6
	Chemicals	3.6	87.2
	Building materials	3.0	90.2
	Electricity	2.4	92.6
	Foods	1.9	94.5
	Textiles	1.3	95.8
	Printing	0.9	96.7
	Glass and pottery	0.8	97.5
	Timber manufactures	0.8	98.3
	Paper	0.7	99.0
	Clothing	0.5	99.5
	Other branches of manufacturing	0.3	99.8
	Leather and footwear	0.2	100.0
1970	Mining	40.6	40.6
	Machinery and electrical engineering	20.8	61.4
	Ferrous metallurgy	14.0	75.4
	Foods	4.4	79.8
	Chemicals	4.2	84.0
	Nonferrous metallurgy	3.5	87.5
	Building materials	3.2	90.7
	Electricity	2.2	92.9
	Clothing	2.1	95.0
	Textiles	1.2	96.2
	Glass and pottery	1.0	97.2
	Timber manufactures	0.8	98.0
	Leather and footwear	0.6	98.6
	Printing	0.6	99.2
	Other manufactures	0.5	99.7
	Paper	0.3	100.0

Source of data: Reprinted with permission from 'Spatial industrial change in Poland since 1945' by R. Wilczewski, T. Lijewski, B. Kortus (1978) in *Industrial Change: International Experience and Public Policy*, edited by F. E. I. Hamilton, published by Longman, London, table 8.6

as a percentage of employment in all industrial activities, are ranked in descending order (Table 2.7, columns 1 and 2).

ii The percentage data are cumulated (Table 2.7, column 3). The first cumulative percentage is the same as the value in column 2, that is, 49.7. The second cumulative percentage is the sum of the first two percentages in column 2, that is, 49.7 + 18.7 = 68.4. The third cumulative percentage is the sum of the first three percentages in column 2, that is, 68.4 + 11.4 = 79.8; and so on.

iii The cumulative percentages are plotted on a graph against the sixteen rank orders. This is done on Figure 2.25. The line joining the point with coordinates 0 percent and rank 0 with the point with coordinates 100 percent and rank 16 represents the most mixed (diversified) industrial employment structure possible in the region. It shows the workforce evenly distributed among the sixteen groups of industrial activity and serves as a yardstick against which to compare the industrial employment structure of one region with another or, as in the present example, against which to compare the industrial employment structure in one region at different times. The closer the actual Lorenz curves lie to the 'line of equality', the greater the regional mix of industry.

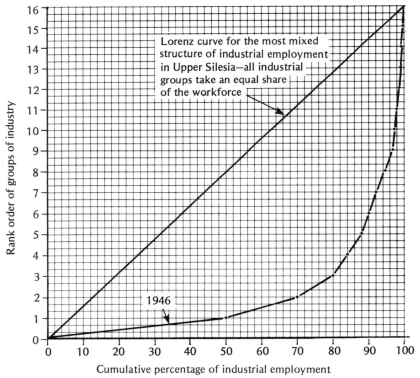

Figure 2.25   Lorenz curves for industrial employment in Upper Silesia, Poland, 1946 and 1970

2 a Using the appropriate data from Table 2.7, construct on Figure 2.25 a Lorenz curve for industrial employment in Upper Silesia in 1970.

b Describe and explain the change in industrial employment structure in Upper Silesia between 1946 and 1970.

c What would be the shape of a Lorenz curve for a region in which just one type of industry was found? (This is the most specialized, or unmixed, industrial employment structure possible.)

Conkling's index of industrial diversification is obtained by dividing the area under the actual Lorenz curve for industrial employment in a region into the area under the Lorenz curve showing the greatest possible mix of industry in the region. A complete mix of industrial employment gives a Conkling index of 1.0 whereas extreme regional specialization gives a Conkling index of 0.0. For Upper Silesia in 1946, Conkling's index is

$$\text{Conkling's index} = \frac{\text{Area under Lorenz curve for Upper Silesia in 1946}}{\text{Area under Lorenz curve describing even mix of industry}}$$

$$= \frac{316 \text{ square units}}{1200 \text{ square units}}$$

$$= 0.26$$

3a Compute Conkling's index for industrial employment in Upper Silesia in 1970. (To find the areas under the Lorenz curves, add up the little squares on the sectional paper, Figure 2.25.)

b Comment on the changes in industrial diversification which have taken place between 1946 and 1970.

Since 1970, attempts have been made to diversify industry in Upper Silesia by investing in growth industries like electrical goods and chemicals and in consumer industries like automobile manufacture.

INDUSTRY IN THE GERMAN DEMOCRATIC REPUBLIC

Prior to 1945, industry in what is now called the German Democratic Republic was concentrated in the South. Rich deposits of lignite near the River Elbe, easy to win by open cast mining but costly to transport, had led to the growth of an electricity-generating industry on a large scale. This, in turn, had attracted chemical industries. The chemical plants used lignite as a raw material and as a source of power, power consumption accounting for over one-third of production costs. Ample supplies of water were to hand in the Elbe-Saale basin.

Horst Kohl has shown that planning in the German Democratic Republic since 1945 has been guided by two factors: firstly, by the desire to reduce the regional imbalance of industry between the North and the South of the country; and secondly, by a chronic shortage of labour supply. In the period between 1950 and 1970 the industrial workforce fell relative to the population which itself dropped by 7 percent. This shrinkage of the labour supply was partly offset by an increase in the number of working women. By 1970, almost half the workforce was female.

Since 1950, major industrial projects have been completed in the dominantly agricultural northern region: oil-refining and petro-chemicals plants at Schwedt, engineering and shipbuilding works at Rostock, steelworks at Eisenhuttenstadt (Figure 2.26). At the same time major industrial

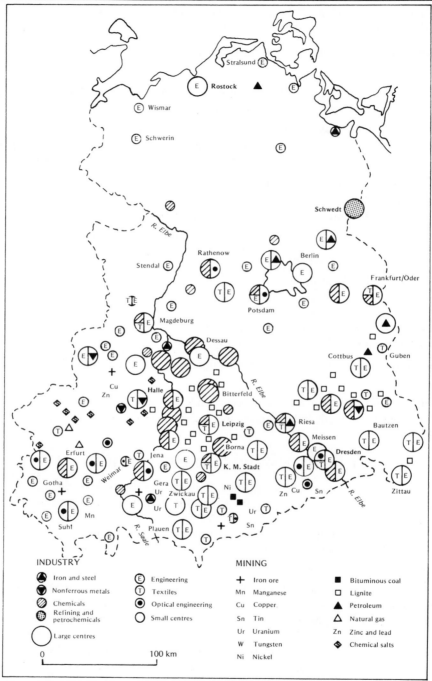

Figure 2.26 The distribution of industry in the German Democratic Republic
Reprinted with permission from *Eastern Europe* by Roy E. H. Mellor (1975), published by
Macmillan, London and Basingstoke, figure 9.1 (Rivers not shown on the original.)

developments have taken place in the southern region: chemical works at Coswig, petro-chemical works at Merseburg, agricultural machinery works at Dresden, and textile machinery works at Karl-Marx-Stadt. The four southern provinces of Halle, Dresden, Leipzig, and Karl-Marx-Stadt still contain just over one-half of the national workforce, though, as in the rest of the country, labour is in short supply.

1 Figure 2.27 depicts the change in the share of the nation's workers in each *Bezirk* (region) between 1956 and 1966. To what extent does the regional pattern of change in the production workforce indicate that attempts to develop industry in the northern region have been successful?

2 In what ways is, and is not, Myrdal's model of economic development of help in explaining the changing pattern of industrial activity in the German Democratic Republic?

3 Suggest why the patterns of industrial location in Poland and in the German Democratic Republic are broadly similar.

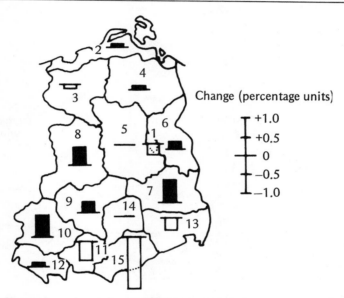

Figure 2.27   Change in regional share of the nation's industrial workers between 1956 and 1966. Regions: 1 Berlin; 2 Rostock; 3 Schwerin; 4 Neubrandenburg; 5 Potsdam; 6 Frankfurt; 7 Cottbus; 8 Magdeburg; 9 Halle; 10 Erfurt; 11 Gera; 12 Suhl; 13 Dresden; 14 Leipzig; 15 Karl-Marx-Stadt

INDUSTRY IN CITIES OF SOCIALIST EUROPE

For a number of reasons, the townscapes of socialist cities are varied. On ideological grounds, the urban environment should be rationally planned with areas of industry and housing separated by open space, but they should not be so far apart as to make commuting time-consuming. One of the earliest examples of this urban land-use plan is Milyutin's concept of a socialist city (Sotsgorod) which he proposed for the planned growth of Stalingrad. In Milyutin's linear city, industry is located on the leeward side of a green belt, on the windward side of which lie the residential areas. Today, Stalingrad, renamed Volgagrad, has a pattern of land use which for the most part follows Milyutin's plan (Figure 2.28a). Similarly, in Siberian cities which have been newly built, industry and housing are kept separate (Figure 2.28b).

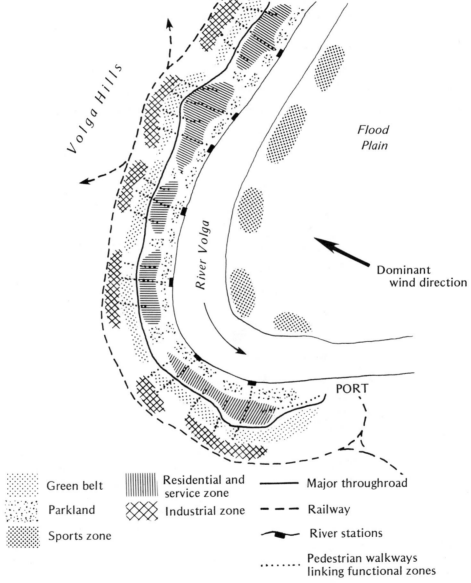

Green belt

Parkland

Sports zone

Residential and service zone

Industrial zone

Major throughroad

Railway

River stations

Pedestrian walkways linking functional zones

Figure 2.28a The linear city: Milyutin's concept for the planned growth of Stalingrad, 1930 Reprinted with permission from *The Socialist City* by R. A. French and F. E. I. Hamilton (1979), published by John Wiley & Sons Ltd., Chichester, figure 1.1. Copyright © 1979 by John Wiley & Sons Ltd

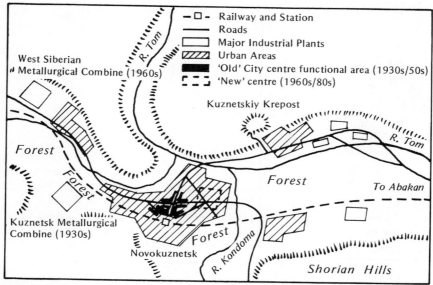

Figure 2.28b Novokuznetsk: an example of a Soviet socialist city in Siberia
Reprinted with permission from *The Socialist City* by R. A. French and F. E. I. Hamilton
(1979), published by John Wiley & Sons Ltd., Chichester, figure 1.3. Copyright © 1979 by
John Wiley & Sons Ltd

Warsaw: A Case Study

Socialist governments in the states of northeastern Europe inherited well-established urban forms.
As a result, planners have been forced to adapt their broad aims of social justice in the city to the
particular urban environments. R. A. French and F. E. I. Hamilton refer to these places as **socialized
cities**. We shall take Warsaw as an example.

Prior to 1918, industry in Warsaw consisted of many small firms scattered throughout the city. The
interwar years witnessed the sprawl of urban development along railway lines, particularly beyond
Wola where industry flourished (Figure 2.29a). In the postwar years, the desire to restore the
city's economy as fast as possible precluded attempts to introduce strict land-use zoning in
accordance with socialist ideology. However, by the mid-1950s, planners were encouraging the
migration of small industrial establishments from sites in the city centre to nearby towns. This
enabled residential areas to be cleared of industry and industrial employment to be provided at
a local level. It also helped to stem the flood of commuters from the suburbs and encourage the
growth of small, fringe towns which had previously stagnated. Factories in the city centre which
had suffered severe damage during the war were demolished and the old industrial districts of
Powisle and Siecle were converted into residential areas with recreational facilities. A strict
zoning of industrial and residential land use was enforced in Wola. More recently, new
industrial estates have been built on the edge of the city (Figure 2.29b).

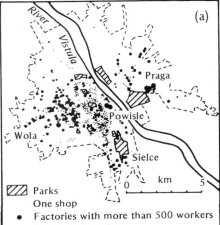

(a)

Parks
One shop
- Factories with more than 500 workers
- 100 to 499 workers
· 50 to 99 workers
--- Edge of built-up area

Figure 2.29   (a) Industry in Warsaw in 1938
(b) Industry in Warsaw in 1970
Reprinted with permission from *The Socialist City*
by R. A. French and F. E. I. Hamilton
(1979), published by John Wiley & Sons Ltd.,
Chichester, figures 13.9 and 13.10. Copyright
© 1979 by John Wiley & Sons Ltd

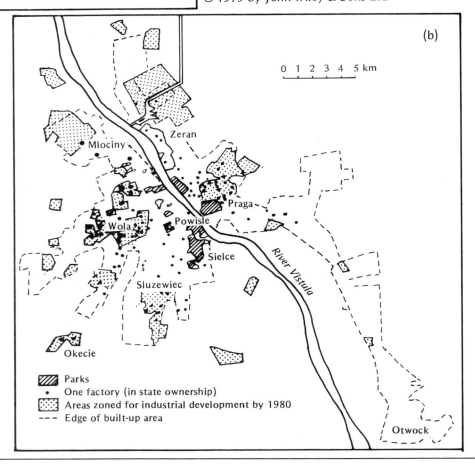

(b)

Parks
· One factory (in state ownership)
Areas zoned for industrial development by 1980
--- Edge of built-up area

Industry in Countries of the Third World

The nations of the Third World have very varied cultures and histories. Nonetheless, many of them share the same broad pattern of industrial development. They all have domestic and craft industries, including pottery, mat-making, and weaving, which date back many centuries. Most of them have passed through a period when their natural resources, including metal ores, nitrates, timber, and rubber, were exploited by European nations. Many of them have, in the last few decades, developed large-scale manufacturing industries, though not so large-scale as in countries of the developed world.

Manufacturing industries in Third World countries fall into three groups.

i.   Manufactures based on preparing and processing the products of agriculture and forestry. These industries include cotton-ginning, the processing of palm oil and groundnut oil, fruit and vegetable canning, sugar-refining, and flour-milling. This group of industries is mainly involved with the initial processing of raw materials prior to export, that is, with what is called export valorization.

ii.   Light industries which use local raw materials to produce consumer goods — textiles, footwear, beer, cigarettes, and so on — for a local market. This obviates the importing of consumer goods from developed nations and is known as import substitution.

iii.   Heavy industries, such as iron- and steel-making, though few Third World countries have developed these in a big way.

The industries found in a particular nation depend on what natural resources are available. But in all nations they are concentrated in a few manufacturing centres. The reason for this lies in history. During colonial rule, one or perhaps two chief ports emerged in each country. These ports served as administrative centres. Many of them became centres of inland transport routes and were later adopted as capital cities. Compared with peripheral parts of countries, the capital cities and their local regions had better transport facilities, were better supplied with energy, and contained the 'better off' portion of the population. For these reasons, capital cities were, during the early stages of industrial growth, attractive areas for setting up industries. To a large extent, the distribution of manufacturing industry today is the same as the distribution which arose at the outset of industrialization. New developments have occurred in established industrial centres. Much of this new development reflects the growing importance of the market which generally coincides with the largest urban centre. In tropical Africa, for example, industrial growth has taken place in areas which provide access to the largest number of people but only in those countries where population size and wealth provide a large enough market for industrial goods. The only major exception to this rule appears to be the Zambian 'copperbelt', though part of the growth of industry in this region can be attributed to the market created by the mining activity.

INDUSTRY IN LATIN AMERICA

Within each country of Latin America, a few manufacturing centres emerged during the early stages of industrial growth. Changes within industries during the interwar period, such as the move from the production of craft goods to the production of consumer durable goods, have strengthened the clustered pattern of industrial development. Consumer durable industries possess a high degree of linkage and are therefore dependent on supplies from other factories. Plants involved in consumer durables manufacture tend to be larger and more efficient than plants producing traditional craft goods. An increasing number of them supply a national, rather than a local, market. The core industrial regions have the biggest sales potential and, in some cases, the lowest costs, and in many other ways remain the most attractive sites for industrial growth (see, for example, Figure 2.30).

Few Latin American governments have encouraged manufacturing industry to spread to peripheral regions from the core areas. Even in Brazil, Chile, Cuba, and Puerto Rico, where regional policies have been pursued through the 'carrot' of financial incentives and development of industrial estates, little spread has taken place. Indeed, Alan Gilbert has observed a tendency for manufacturing industry within the countries of Latin America to become even more concentrated than it was (Table 2.8). This tendency to increased concentration has contributed to the marked predominance of one city, the primate city, in each country (see *Settlements*).

Figure 2.30   Latin America: the distribution of electricity-generating capacity
Reprinted with permission from P. R. Odell and D. A. Preston (1973), *Economies and Societies in Latin America: A Geographical Interpretation*, published by John Wiley & Sons Ltd., Chichester, figure 8.2. Copyright © 1973 John Wiley & Sons Ltd

Table 2.8

Country	Largest city	Period of change	Percent share in national manufacturing employment		Change in percent share in national manufacturing employment	Index of primacy
			Start of period	End of period		
Argentina	Buenos Aires	1935 to 1965	25.2	39.9		7.9
Brazil	São Paulo	1940 to 1968	35.0	50.2		1.3
Chile	Santiago	1952 to 1954	49.3	60.3		4.5
Colombia	Bogota	1945 to 1967	16.7	25.3		2.2
Ecuador	Guayaquil	1950 to 1965	38.9	38.8		1.2
Mexico	Mexico City	1930 to 1965	27.5	46.1		7.4
Peru	Lima	1940 to 1963	13.7	70.2		8.0

Source of data: percentage shares in national manufacturing employment is from *Latin American Development* by A. Gilbert, published by Penguin Books 1977, table 8; index of primacy is from *Settlements*, table 1.10.

1 Using the information given in Table 2.8, compare the percent change in national manufacturing employment with the index of primacy and describe in what way the two are related. (The index of primacy is the ratio between the population of the largest city in a nation and the population of the second largest city. An index value of 1.0 would mean that the second city was as large as the first. An index of 2.0 would mean that the second city was half as large as the first, and so forth. In general, the higher the index, the more dominant is the largest city.)

INDUSTRIAL MOVEMENT IN NIGERIA

In the countries of tropical Africa, as in the countries of Latin America (p. 127), industry is concentrated in ports and primate cities (Table 2.9). Even in land-locked nations, industry tends to be found at break-of-bulk sites. In Nigeria, where the volume of imports and the volume of partially processed raw material exports have increased during the period of industrial expansion, just over one-third of all manufacturing plants were located in Lagos and the old colonial province in 1960 (Figure 2.31). To help overcome the regional

Table 2.9

Country	Primate city	Percent share of national manufacturing employment
Sierra Leone	Freetown	75.0
Senegal	Dakar	81.5
Gambia	Banjul	100.0
Ghana	Accra	30.4
Nigeria	Lagos	35.0
Ivory Coast	Abidjan	62.5
Liberia	Monrovia	100.0

Source of data: 'Manufacturing and the geography of development in tropical Africa' by A. L. Mabogunje, *Economic Geography*, 49 (1973), 1–20

imbalance of industry, the Nigerian government has constructed industrial estates which come fully equipped with modern amenities. The government has not forced entrepreneurs to locate in peripheral areas and the most popular industrial estates lie close to established industrial centres. However, in 1967, twelve state capitals were created, in all of which transport, power supply, and other services have been improved. These expanding administrative centres have attracted some manufacturing plants. So, in Nigeria, government influence on the location of industry has been indirect.

J. Okezie and C. Onyemelukwe have studied the relocation of industry in Nigeria during the period 1960 to 1970. They identified a number of factors which played a part in inducing firms to move. First of all, industries that wished to expand but were constrained from doing so at or near their existing site, because of lack of space or high costs of land and labour, were forced to look to remoter regions to set up a branch plant. Secondly, some industries on the look out for new sites were drawn to areas where transport and power supply facilities had been improved. Thirdly, some industries avoided areas where war and natural hazards were likely to occur. Fourthly, some entrepreneurs sought personal prestige by relocating in their home territories. Fifthly, govenment activity had a little influence on the relocation of some industries but, because of a low level of technology and a large amount of foreign investment, not much.

1   Examine Table 2.10. Do the reasons for the relocation of main plants and the relocation of branch plants differ? Account for the differences and similarities.

2   Would the pattern of movement shown on Figure 2.32 be predicted by a simple gravity model like the one described on p. 101?

3   Identify and describe some of the processes recognized by Myrdal and by Friedmann which seem to have been at work in the industrial and economic development of Nigeria.

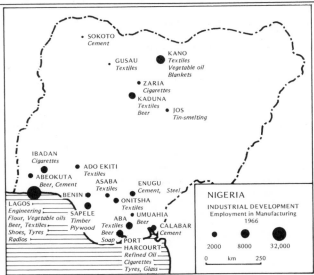

Figure 2.31   Industrial development in Nigeria: centres with more than 1000 employees in manufacturing, 1966
Reprinted with permission from *The Geography of Tropical Development* (second edition) by A. M. O'Connor (1978), published by Pergamon Press Ltd., Oxford, map 5.4

Table 2.10 Type and explanation of migration of industrial units in Nigeria, 1974

Total number of moves: 42	Type of movement involved			
	Main plant relocation		Branch plant relocation	
	19		23	
Reasons for migration:	Primary	Secondary	Primary	Secondary
Need to expand in branch plants			13	4
Lack of space for *in situ* expansion	1	3	1	3
Better production prospects in a new place	1	2	3	2
Desire to be nearer home	2	1	4	5
Influence of government	8	2	2	3
Greater safety for plant in a new place	7	2		

Source of data: Field Survey and Questionnaire, May—July, 1974
Reprinted with permission from 'Industrial Movement in West Africa: the Nigerian case' by
J. Okezie and C. Onyemelukwe (1978) in *Industrial Change: International Experience and
Public Policy*, edited by F. E. I. Hamilton, published by Longman, London, table 10.2

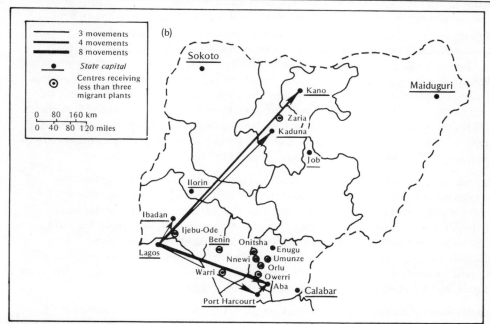

Figure 2.32 Industrial migration in Nigeria, 1960 to 1970
Reprinted with permission from 'Industrial migration in West Africa: the Nigerian case' by
J. Okezie and C. Onyemelukwe (1978) in *Industrial Change: International Experience and
Public Policy*, edited by F. E. Ian Hamilton, published by Longman, London, figure 10.3

Industry in Urban Areas

URBAN INDUSTRIAL LANDSCAPES

The pattern of industrial location in urban areas is moulded by the same forces that mould the pattern of industrial location in rural areas, namely, land, labour, materials, markets, and the whims of entrepreneurs. But cities, in particular large ones, have special industrial opportunities. They have administrative, educational, and commercial functions, a large labour force, and a large and varied market. These special attributes contribute to a distinctive industrial structure — clothing, luxury goods, office and business equipment, scientific instruments, vehicles, electrical equipment, light chemicals, pharmaceuticals, furniture, and food are all made. These industries are not haphazardly dotted around a city. As F. E. I. Hamilton explained, differences in the cost of land, labour, and transport from one urban site to another sort and sift them into a number of zones. Central areas contain, among other industries, instrument and tool manufacturers and printers, who need to draw on a skilled labour force from the entire city; clothing and office equipment makers, who need access to the enormous sales possibilities offered by the central business district; and newspapers, who find that ready access to the whole urban market offsets the high prices paid for land. Small-scale agglomeration of allied industries is a commonplace in central city areas — industrial quarters are found. An example is the East End clothing quarter of London covering a square mile of Whitechapel and Spitalfields with firms specializing in buttons and buckles, buttonholing and button-covering, belts and shoulder pads, pleating and embroidery, and sewing machine repairs. Other examples from London are the Camden Town instruments' area and the Hackney shoes' quarter.

Larger industries — makers of food and electrical goods, engineering firms, many other light industries, and port industries — need cheaper land, a good site for assembling materials and distributing products, and access to a skilled or semi-skilled labour force. These requirements are met along transport routes running out of, and round, the more suburban parts of a city. Industries needing large areas of land for assembly-line production, for stores and dumping waste, or which are dangerous or produce obnoxious smells or a lot of noise — vehicle manufacturers, heavy engineering, oil-refining, heavy chemicals, metal-making, and paper-making — are found on the edge of a city.

This pattern of urban industry is also found in the United States. An examination of 44 American cities made by L. K. Loewenstein revealed that, overall, manufacturing activity follows the lines of railways, roads, and rivers in the outer parts of cities. Many manufacturing industries require single-storey buildings and a lot of space; these are not available in a city centre. In any case, many industries cannot afford the high rents which central sites command. But some can. Industries orientated towards skilled labour supplies and urban markets do tend to occupy sites near the central city core. From a central location, urban market-orientated industries like newspaper publishing and commercial printing, whose products are highly competitive in price, can serve their markets with the minimum of distribution costs. And industries like cabinet- and furniture-making and medical equipment-producing, which call for a skilled labour force, can draw on workers from the entire city. Integrated industries, such as vehicle assembly, cluster along rail routes, especially at sites with good road links. Industries which handle bulky raw materials, need a lot of land, create noise, or produce noisome smells — these include large, basic processing industries like oil-refining and steelworks — tend to cluster along transport routes. Large, new plants are found on the urban fringe in areas where land is cheap and available in large lots alongside major roads.

1 Attempt to explain the differences and similiarities in the distribution of manufacturing industry in London (Figure 2.33a) and Paris (Figure 2.33b).

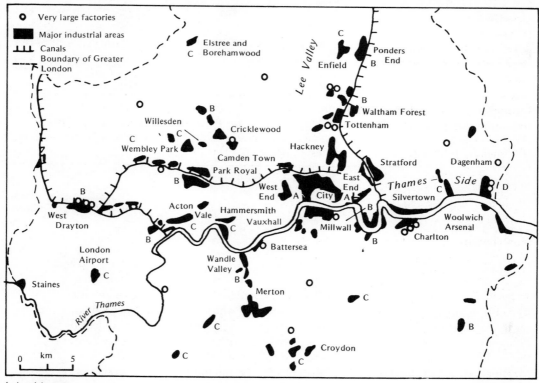

Industrial types:
A. City centre B. Port C. Light D. Heavy

Figure 2.33a Major industrial areas in Greater London
Based on the work of J. E. Martin

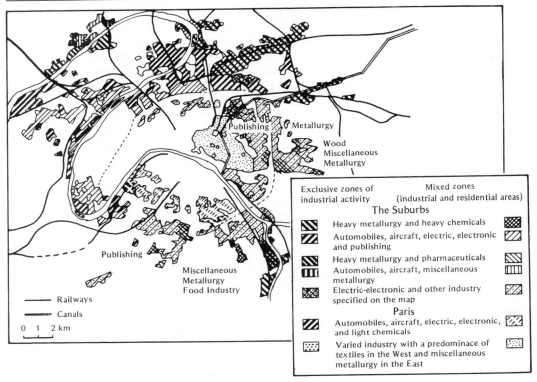

Railways

Canals

0 1 2 km

Figure 2.33b   The distribution of industry in Paris
Reprinted with permission from 'Industrial activity in the Parisian agglomeration' by J. Bastié
(1975) in *Locational Dynamics of Manufacturing Activity*, edited by L. Collins and D. F. Walker,
published by John Wiley & Sons, Chichester, figure 10.1

New Towns

New Towns differ from other settlements in that planners start with a clean slate and are
free to locate activities, including industry, wherever they choose. Milton Keynes was
designated a New Town in 1967 and had a target population of 250,000. To administer
its construction, a development corporation was set up which considered five proposals
for the location of industry (Figure 2.34). Given the size, shape, and residential density
(8 houses per acre) of the area, calculations were made of the total distance covered by
the workforce in getting to work, the length of roads required, the cost of road
construction, and the 'convenience' of each plan to the community. Plan 5, dispersed
industrial location (Figure 2.34), made the smallest hole in the Development Corporation's
pocket and, along with plan 1, was the most 'convenient' plan for the community at large.
Figure 2.35 shows how industrial sites have been spread around the urban area, pockets of
light industry being located within residential areas for the benefit of female workers.

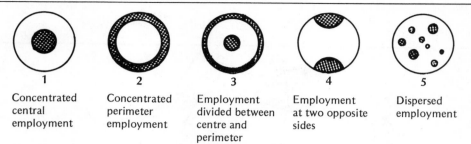

1	2	3	4	5
Concentrated central employment	Concentrated perimeter employment	Employment divided between centre and perimeter	Employment at two opposite sides	Dispersed employment

Figure 2.34   Five plans for the layout of Milton Keynes
Reprinted with permission from *New Towns: The British Experience* edited by Hazel Evans (1972), published by Charles Knight and Co. Ltd., London, figure 17

INDUSTRIAL TOWNS

The central-place hierarchies developed by Christaller and Lösch predict an orderly pattern in the provision of services within a set of settlements. In general, the larger the settlement, the greater the number and wider the range of services it offers. As we have seen, large cities do have a wide range and variety of manufacturing activities. But some towns seem to go against the general trend and specialize in the production of one good. The colliery towns in northern England and the cotton towns of the Appalachians are cases in point.

G. Alexandersson made an extensive study of the occupational structure of 864 cities with populations in excess of 10,000 in the United States in 1950. He found that 23 out of 36 groups of industrial activity were represented in all cities. The remaining 13 groups of industrial activity were found in just some of the cities. All service industries were ubiquitous, that is, found in all cities. Three groups of manufacturing industry — construction, printing and publishing, and food-processing — were found in all cities but all other groups of manufacturing industry were to some extent sporadic, that is, found in just some of the cities. The motor car industry, for instance, was highly sporadic, being found in a very few towns (cf. Figure 1.46).

Extreme cases of industrial specialization are found in towns owned by companies. Various sorts of **company towns** exist. In remote areas, where resources are exploited, towns are found at sites of mineral extraction or, in the cases of fishing, lumbering, and oil-drilling, at convenient points of collection. The atmosphere of an extractive company town, a sawmill settlement on the Mississippi-Alabama border early in this century, is evoked by William Faulkner in his book *Light in August*:

> All the men in the village worked in the mill or for it. It was cutting pine. It had been there seven years and in seven years more it would destroy all the timber within its reach. Then some of the machinery and most of the men who ran it and existed because of and for it would be loaded onto freight cars and moved away . . . leaving . . . a stumppocked scene of profound and peaceful desolation. . . . Then the hamlet which in its best day had borne no name listed on Post Office Department annals would not now even be remembered by the hookwormridden heirs-at-large who pulled the buildings down and burned them in cook-stoves and winter grates.

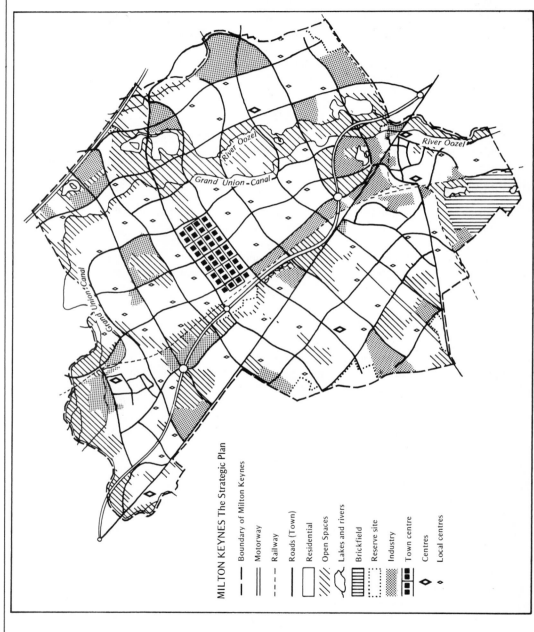

MILTON KEYNES The Strategic Plan

Boundary of Milton Keynes
Motorway
Railway
Roads (Town)
Residential
Open Spaces
Lakes and rivers
Brickfield
Reserve site
Industry
Town centre
Centres
Local centres

Figure 2.35 Milton Keynes: the stategic plan
Reprinted with permission of the Milton Keynes Development Corporation

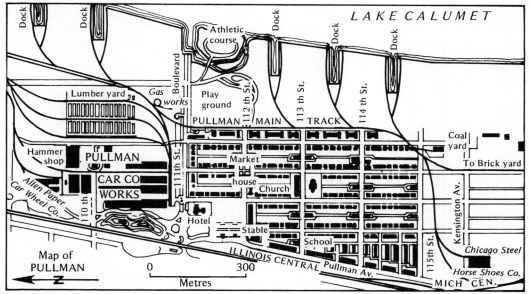

Figure 2.36 Pullman, Illinois, a model town created by George M. Pullman
Source: *Harper's Monthly*, 1885

Another sort of company town is the **factory town**. An example is Pullman, Illinois, a model town created by George M. Pullman for the production of sleeping cars and a 'new era of labor' (Figure 2.36).

There are a large number of company towns based on transport. One of the longest lived of these is Goole in Humberside which started life under the aegis of a canal company and was later supported by a railway company. Ellesmere Port, Runcorn, and Stourport also grew up as canal company ports.

1 The degree to which the industrial activity of a town is specialized may be gauged by constructing a Lorenz curve of employment data. A Lorenz curve for Stoke-on-Trent which specializes in pottery-making, is shown on Figure 2.37.

a Using the data listed in Table 2.11, construct Lorenz curves for one of these towns: Corby, Coventry, Newport, Portsmouth, Telford.

b Describe and attempt to explain your results.

INDUSTRIAL CHANGE IN CITIES

Industrial patterns within cities are not static. For example, in Portsmouth the employment structure was originally dominated by the Royal Naval Dockyard. Even as late as 1939, 60 percent of the male labour force in manufacturing industry was employed by the government. The dockyard (see *Settlements*, p. 156) acted as a magnet: industries congregated close to the dockyard and the high-density residential areas of Kingston, Buckland, and Fratton. Even with

138

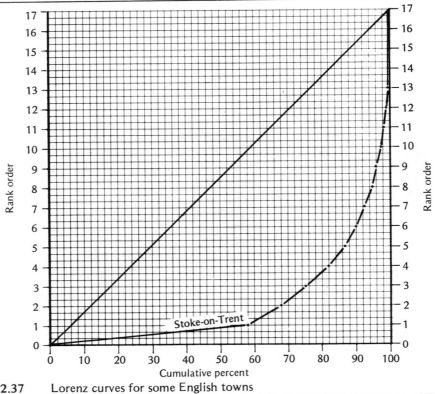

Figure 2.37 Lorenz curves for some English towns

Table 2.11 Manufacturing employment in some English towns (thousands)

Industrial group	Coventry	Newport	Corby	Portsmouth	Telford
Food, drink, and tobacco	1.6	1.5	0.8	1.9	0.2
Coal and petroleum products	0.1	Nil	Nil	Nil	Nil
Chemicals and allied industries	0.1	1.7	0.2	0.5	0.3
Metal manufacture	1.0	17.1	12.2	0.1	8.2
Mechanical engineering	21.2	3.3	0.3	3.1	9.3
Instrument engineering	0.1	Nil	Nil	0.1	Nil
Electrical engineering	19.1	3.6	0.2	5.8	3.2
Shipbuilding and marine engineering	Nil	Nil	0.1	10.4	Nil
Vehicles	70.2	0.2	0.3	2.7	15.4
Other metal goods	6.9	0.5	0.1	0.6	5.1
Textiles	5.8	0.1	0.3	0.1	0.6
Clothing and footwear	0.2	0.9	1.3	1.5	1.8
Bricks, pottery, glass, cement, etc.	0.8	0.4	0.2	0.1	1.0
Timber, furniture, etc.	1.1	0.3	Nil	0.9	0.3
Paper, printing, and publishing	1.5	1.7	Nil	2.2	1.2
Other manufacturing industries	1.2	0.5	Nil	2.6	2.6
Totals	130.9	31.8	15.9	32.5	49.2

the development of tramways, which hastened urban sprawl on Portsea Island, the pattern of industrial location did not change, partly because industrial links with the dockyard were very strong. It was not until 1945 that significant changes took place, with movement of industry from the older parts of the city. The industries that have remained in the old core area, with the exception of the Brickwood's brewery, tend to be small-scale enterprises like printing (Figure 2.38).

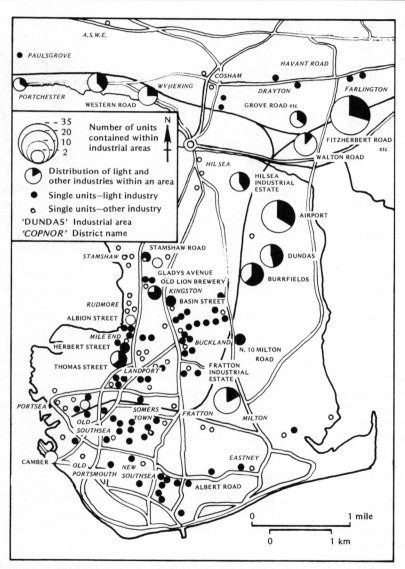

Figure 2.38   The distribution of manufacturing industry in Portsmouth, 1971
Reprinted with permission from 'The Development of Manufacturing in Portsmouth' by Robert Jackson (1974) in *Portsmouth Geographical Essays*, published by Portsmouth Polytechnic, figure 5

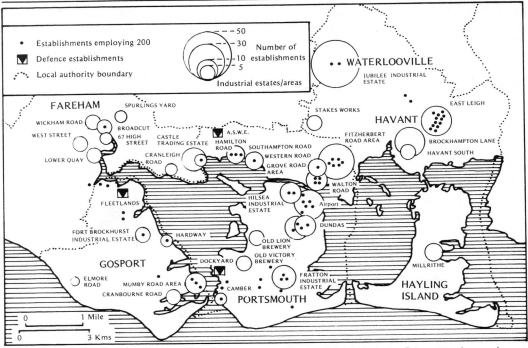

Figure 2.39   Industrial estates and large industrial establishments in the Portsmouth area in 1971
Reprinted with permission from 'The Development of Manufacturing in Portsmouth' by Robert Jackson (1974) in *Portsmouth Geographical Essays*, published by Portsmouth Polytechnic, figure 6

In addition to relocation of industry within Portsmouth, there has been an influx of new industry, including pharmaceuticals and electrical goods, to peripheral sites. The availability of land has been one of the most important factors influencing the new distribution of industry, reclaimed land in the northern part of Portsmouth harbour offering attractive industrial sites. The modern, purpose-built industrial estates are not mixed with residential areas; this segregation between places of work and homes was not characteristic of nineteenth-century Portsmouth (Figure 2.39).

Change in the Number of Manufacturing Establishments

The case of Portsmouth reveals two important features of industrial change in cities: firstly, an overall change in the number of manufacturing establishments brought about by the 'births' and 'deaths' of firms; and secondly, a relocation of firms which usually involves a move out of central sites to the suburbs.

Y. S. Cohen and B. J. L. Berry have made a study of the relationship between the change in the number of manufacturing establishments and city size for 3102 counties in the United States between 1952 and 1962. Some of their findings are summarized in Table 2.12a. The data suggest that growth in the number of manufacturing establishments has been associated with large cities.

Table 2.12a Observed frequencies

Change in the number of manufacturing establishments	Size of city (population of city core)			Row totals
	Small (Less than 25,000)	Medium (25,000 to 100,000)	Large (More than 100,000)	
Decline	397	263	209	869
Growth	671	756	806	2233
Column totals	1068	1019	1015	3102

To discover if this apparent relationship is borne out by statistical analysis, a test may be run which compares the observed frequencies of cities falling in each of the six growth-decline—city-size categories shown in Table 2.12a with the frequencies which would be expected to exist if there were *no* relationship between the change in the number of manufacturing establishments and city size. The test, called the chi-squared test, runs like this:

Step one   Computing expected cell frequencies
The expected frequency of small cities showing a decline in the number of manufacturing establishments is computed by multiplying the total number of small cities (this is 1068, the total of column 1 in Table 2.12a) by the total number of cities of all sizes which have suffered a decline in the number of manufacturing establishments (this is 869, the total of row 1 in Table 2.12a), and dividing the result by the total number of cities in the study (this is 3102). We have

Expected number of
small cities showing
a decline in the number = $\dfrac{1068 \times 869}{3102}$ = 299.2.
of manufacturing
establishments

The remaining five expected frequencies are computed in like manner as shown in Table 2.12b.

Table 2.12b Expected frequencies

Change in the number of manufacturing establishments	Size of city (population of city core)			Row totals
	Small (Less than 25,000)	Medium (25,000 to 100,000)	Large (More than 100,000)	
Decline	$\dfrac{1068 \times 869}{3102}$ = 299.2	$\dfrac{1019 \times 869}{3102}$ = 285.5	$\dfrac{1015 \times 869}{3102}$ = 284.3	869
Growth	$\dfrac{1068 \times 2233}{3102}$ = 768.8	$\dfrac{1019 \times 2233}{3102}$ = 733.5	$\dfrac{1015 \times 2233}{3102}$ = 730.7	2233
Column totals	1068	1019	1015	3102

Step two   Calculating the chi-squared statistic

The next step is, for each of the six city-size—growth-decline combinations, to subtract the expected frequency from the observed frequency, square the result, then divide by the expected frequency. For small cities showing a decline in the number of manufacturing establishments we have

$$\frac{(\text{Observed frequency} - \text{Expected frequency})^2}{\text{Expected frequency}} = \frac{(397 - 299.2)^2}{299.2}$$

$$= 32.0.$$

This procedure is carried out for all city-size — growth-decline combinations and the resulting six values added up (Table 2.12c). The sum of the six values is the observed chi-squared value, written χ^2_{obs}, and in the example is 74.6.

Table 2.12c   Calculating the observed chi-squared value, χ^2_{obs}

Change in the number of manufacturing establishments	Size of city (population of city core)			Row totals
	Small (Less than 25,000)	Medium (25,000 to 100,000)	Large (More than 100,000)	
Decline	$\frac{(397 - 299.2)^2}{299.2} = 32.0$	$\frac{(263 - 285.5)^2}{285.5} = 1.8$	$\frac{(209 - 284.3)^2}{284.3} = 19.9$	53.7
Growth	$\frac{(671 - 768.8)^2}{768.8} = 12.4$	$\frac{(756 - 733.5)^2}{733.5} = 0.7$	$\frac{(806 - 730.7)^2}{730.7} = 7.8$	20.9
Column totals	44.4	2.5	27.7	$\chi^2_{obs} = 74.6*$

*Notice that the observed chi-squared value is given by either the sum of the row totals *or* the sum of the column totals in this table.

Step three   Putting the result to the test

To decide whether to accept or to reject the hypothesis that city size and change in number of manufacturing establishments are *not* related (this is termed a null hypothesis since it assumes *no* relationship), the observed chi-squared value is compared with what is called a critical chi-squared value which may be read off from Figure 2.40. But to make the comparison we need to calculate what are termed the degrees of freedom; these are defined for a table as the product of the number of rows less one and the number of columns less one. In the example, with two rows and three columns, the degrees of freedom are $(2 - 1)(3 - 1) = 2$. We also need to define what level of statistical significance we are prepared to accept in affirming or rejecting the null hypothesis. We shall take the 99 percent significance level, the line for which on Figure 2.40 intersects the two-degrees-of-freedom line at a critical chi-squared value, χ^2_{crit}, of 9.2. The observed chi-squared value, 74.6, is much greater than the critical chi-squared value and so we may reject the null hypothesis that change in the number of manufacturing establishments and city size are not related. We may conclude therefore, that, in 99 cases out of 100, change in the number of manufacturing establishments between 1952 and 1962 in the 3102 cities studied in the United States *was* related to city size by more than a chance association.

1   What geographical processes might explain this relationship?

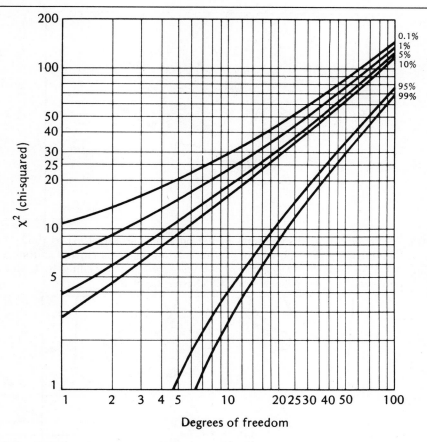

Figure 2.40 Graph for use in the chi-squared test

The numbers on the right of the diagram refer to the lines and are probabilities. For example, the value 0.1 percent means that the association measured by the chi-squared value will occur by chance once in a thousand times; the significance level is thus 99.9 percent. The probability value of 1 percent corresponds to a significance value of 99 percent, the probability value of 5 percent to a 95 percent significance level; and so forth

Reprinted with permission from *Statistical Methods and the Geographer* by S. Gregory (1978), published by Longman, London and New York

The Relocation of Industry within Cities

The movement of industry from city centre to suburbs has taken place in many developed countries. In Paris, migration to the suburbs has been encouraged by the government and improvement in transport links. In Toronto, there has been considerable relocation of industry within the metropolitan area to parts of the city which have in recent years experienced a rapid growth of population (Figures 2.41a and b).

An undesirable result of industrial relocation has been the creation of inner city areas lacking employment opportunities. In the United States, suburban growth coupled with decentralization

144

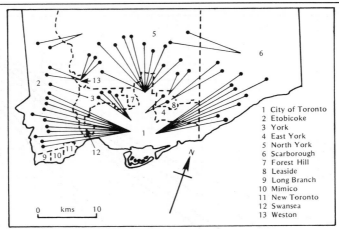

Figure 2.41a The relocation of some firms in metropolitan Toronto, 1961 to 1962
Reprinted with permission from 'A procedure for forecasting changes in manufacturing industry'
by L. Collins (1975) in *Locational Dynamics of Manufacturing Activity* edited by L. Collins and
D. F. Walker, published by John Wiley & Sons, Chichester, figure 8.4

Figure 2.41b Areas in metropolitan Toronto with a population growth rate in the period 1951
to 1961 in excess of 300 percent (shaded)
Adapted from M. Yeates and B. H. Garner (1976) *The North American City*, published by Harper
& Row, New York, figure 10.4

of industry has left many cities with socially and economically depressed cores. In these cores,
ghettos have grown up. In the light of studies carried out on ghettos and their problems, such as
the Kerner Commission of 1968 on civil disorders, more financial aid is now given to urban areas.

Similar problems of inner city areas have been encountered in the United Kingdom. In the
Greater London Council area, employment declined during the period 1966 to 1973, 70
percent of the jobs being lost in manufacturing industries. As a result, a number of people moved
out of the central area. But not all workers were in a position to move out and the result has been
high unemployment in inner boroughs.

2a Complete Table 2.13 by calculating, for both 1951 and 1977, the population of each borough expressed as a percentage of the total population of Greater London, and then calculating the change in percentage share of population in each borough between 1951 and 1977.

b Construct choropleth maps of change in share of population (Figure 2.42a) and unemployment (Figure 2.42b).

c Construct a scatter-graph (Figure 2.43) of unemployment plotted against change in share of population. Put in a line of best fit.

d Examine your maps and your scatter-graph and describe the relationship between unemployment and population change in Greater London.

The British government commissioned in 1976 a number of studies in inner city areas including Lambeth (London), Granby and Edge Hill (Liverpool), and Small Heath (Birmingham). These studies led to the passing of the Inner Urban Areas Act of 1978 which is an attempt to bring industry back to city centres.

Table 2.13 Population and unemployment data for the Greater London boroughs

Reference number	Borough	Population 1951 (Thousands)	Population 1951 Percentage share of total	Population 1971 (Thousands)	Population 1971 Percentage share of total	Change in share of total population, 1951 to 1971 (percentage units)	Unemployment, 1977 (percent)
1	City of London	5	0.1	5	0.1	0	2.6
2	Barking	189	2.3	153	2.2	−0.1	2.7
3	Barnet	320	3.9	292	4.2	+0.3	2.0
4	Bexley	205	2.5	215	3.1	+0.6	1.8
5	Brent	311	3.8	256	3.7	−0.1	3.4
6	Bromley	268	3.3	293	4.2	+0.9	2.5
7	Camden	258	3.2	189	2.7	−0.5	4.1
8	Croydon	310	3.8	322	4.6	+0.8	2.3
9	Ealing	311	3.8	292	4.2	+0.4	2.7
10	Enfield	288	3.5	260	3.7	+0.2	1.9
11	Greenwich	236	2.9	206	3.0	+0.1	3.2
12	Hackney	265	3.2	194	2.8	−0.4	5.4
13	Hammersmith	241	2.9	164	2.4	−0.5	4.4
14	Haringey	277	3.4	228	3.3	−0.1	3.2
15	Harrow	219	2.7	199	2.9	+0.2	1.8
16	Havering	192	2.3	240	3.4	+1.1	2.0
17	Hillingdon	210	2.6	229	3.3	+0.7	1.9
18	Hounslow	211	2.6	200	2.9	+0.3	2.5
19	Islington	271	3.3	168	2.4	−0.9	4.6
20	Kensington and Chelsea	219	2.7	159	2.3	−0.4	3.5
21	Kingston upon Thames	147	1.8	136	2.0	+0.2	1.5
22	Lambeth	347	4.2	283	4.1	−0.1	4.5
23	Lewisham	303		244			3.4
24	Merton	200		166			2.2
25	Newham	294		230			4.4
26	Redbridge	257		229			2.6
27	Richmond upon Thames	188		165			1.9
28	Southwark	338		224			4.5
29	Sutton	176		167			1.9
30	Tower Hamlets	231		150			5.7
31	Waltham Forest	275		222			2.4
32	Wandsworth	331		278			3.7
33	Westminster, City of	300		210			3.8
	Greater London	**8197**		**6968**			

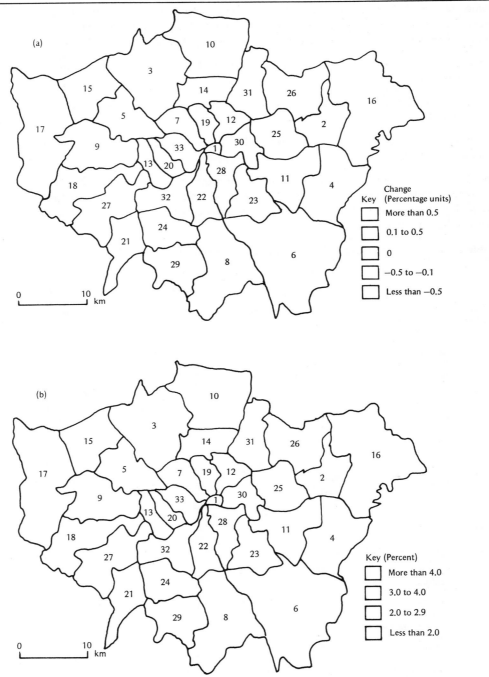

Figure 2.42 (a) Change in share of population by borough, 1951 to 1977
(b) Unemployment, 1977
Boroughs may be identified from Table 2.13.

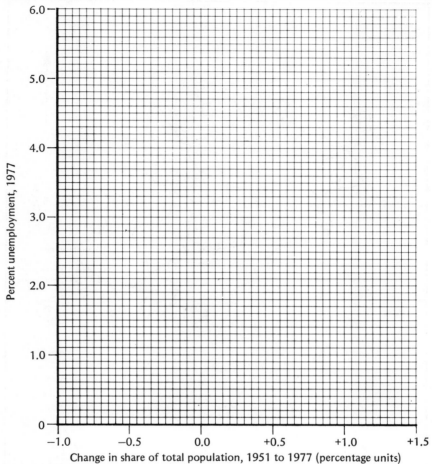

Figure 2.43 A scatter-graph on which percent unemployment in 1977 is plotted against change in share of total population between 1951 and 1977 in the boroughs of Greater London

Small Heath, Birmingham: A Case Study

Small Heath lies two kilometres to the east-south-east of Birmingham's city centre (Figure 2.44). It mostly consists of late Victorian houses. Its decline over the last three decades has been dramatic. Since 1951, its population has fallen by one-third and many jobs have been lost. Unemployment is now running at just over 20 percent. Skilled and semi-skilled workers have tended to move out leaving a body of unskilled workers, many of whom are coloured.

The loss of manufacturing employment in Small Heath has resulted largely from a high death-rate of firms. Many large establishments have, for a variety of reasons, including their outmoded equipment and the general economic recession, closed down. They have not yet been replaced by new firms. However, as a result of the Inner Urban Areas Act of 1978, £115 million has been made available to provide loans for construction and modernization of buildings, the installation of services, and so on in areas like Small Heath.

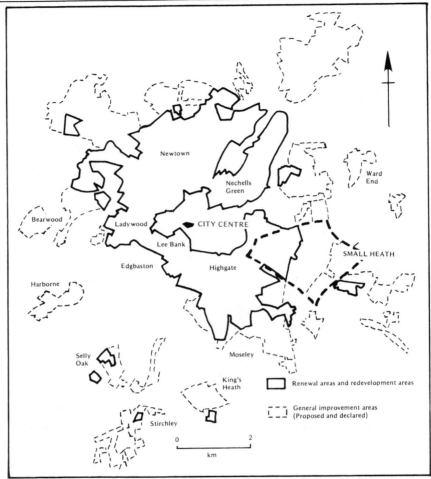

Figure 2.44 The location of Small Heath
Source: Based on the work of P. N. Jones

3 Read the following extract from *Unequal City*, a report of the Birmingham Inner Area Study by consultants made on behalf of the Department of Environment. To what extent are the consultants' recommendations similar to government policy on assistance to industry in peripheral regions in 1980?

Recommendations for employment

We advocate a policy which operates on the preservation or creation of jobs in or readily accessible from the area rather than on their movement, which subsidises employment directly as well as through subsidising investment, and, which takes account of the area's vulnerability to the trade cycle. In view of constraints on resources it should also fit in as part of national policy. In our opinion these requirements might be met by a subsidy on all new jobs in the manufacturing sector. This would be paid in the job's first year at a declared rate, and over the next six to nine years at a tapering rate. The declared rate for new jobs would be graduated annually by central government in respect of conditions of unemployment nationally and at the regional level by (in this case) the West Midland County Council in respect of spatial discrimination. We realise that there may be practical difficulties in implementing this proposal and that it may also be more applicable to other inner areas than to Small Heath. However, if accepted, the implementation of this policy would be a matter of urgency.

We recommend that the City, together with the County, in planning employment should include in their consideration:

i. A vigilant attention to the balance between housing and employment such that skilled manual workers are not lost unnecessarily to the city's workforce.
ii. The development of modern industrial estates.
iii. The rethinking of planning and zoning policy, including the treatment of non-conforming uses.
iv. The making available of grants for the rehabilitation of industrial premises and for their upgrading to meet safety and environmental protection standards.
v. A greater capacity to respond to the needs of industry.

4 According to the Smaller Firms' Branch of the Confederation of British Industry, two big obstacles hinder the expansion of manufacturing industries in inner city areas. These obstacles are existing planning procedures and property rate levels.

a Explain this statement.

b How is government policy on inner city areas introduced in 1980 attempting to overcome these obstacles? (Read the extract from *The Times* on the following page before you attempt this question.)

Enterprise zones to head drive for new jobs

By Baron Phillips

Some of the most rundown inner city areas in the country received a boost yesterday when Mrs Margaret Thatcher announced in the Commons that seven were to be designated as enterprise zones.

These zones – first announced by Sir Geoffrey Howe, the Chancellor of the Exchequer, in his Budget speech in March – are aimed at stimulating private investment in deprived areas.

Detailed negotiations still have to be undertaken with the local authorities and legislation will have to go through Parliament. However, the aim is for the first seven zones to be formally established next spring.

Mrs Thatcher said in the House of Commons yesterday that additional zones would be designated later. One of these would be in the Midlands, in Wolverhampton, Dudley or Corby.

Under the Government's proposals, each zone of up to

The seven zones

- **Inner area of Belfast**
- **Lower Swansea Valley**
- **Clydebank**
- **Parts of Newcastle and Gateshead**
- **Speke (Liverpool)**
- **Parts of Salford docks and Trafford Park, Manchester**
- **Isle of Dogs**

500 acres will be granted concessions and exemptions on rates, planning restrictions, capital allowances and priority over Customs clearances.

The biggest single cash concession the Government is making is on rates. Any company setting up in an enterprise zone, including companies already set up in the areas, will be exempt from paying rates for the 10-year duration of the enterprise plan.

But no Government money will be pumped directly into these zones. Instead, money will be paid by the Government to the local authorities to cover the loss of rates.

This is expected to cost around £10m in a full year, although the figure will grow over the 10 years. Enhanced capital allowances on investment in industrial and commercial buildings is likely to cost the state £20m in lost revenue in a full year.

Owners of land and buildings in the zones will be exempt from paying development land tax on property which they sell for development.

To encourage industry and commerce to move into these areas, planning controls and procedures will be kept to a minimum. There will be no relaxation of regulations on health, safety and the control of pollution, but the administration of these will be streamlined.

Sir Horace Cutler, leader of the Greater London Council, said last night that the establishment of the Isle of Dogs enterprise zone would provide a "tremendous boost" for dockland.

The setting up of an enterprise zone for Clydebank met with an enthusiastic response from the local council and the Scottish Development Agency.

Mr Lewis Robertson, chief executive of the Scottish Development Agency, said in Glasgow last night: "The enterprise zone at Clydebank will be a considerable bonus to the task force which has recently set up in the town."

In the Greater Manchester area the decision was welcomed by both the North West Industrial Development Association and the city's chamber of commerce, both of which have argued strongly for a second zone in the region.

Reprinted by permission of *The Times*, 30 July 1980

Summary

Within countries, there is usually an imbalance of industrial and economic activity between one part and another, between core regions and peripheral regions. This regional imbalance is found in developed and developing, capitalist and communist countries alike.

The pattern of change in the economic fortunes of regions in developing countries has been subjected to theoretical analysis by Myrdal and by Friedmann. The theoretical frameworks of these economists are also applicable to the study of economic and industrial growth in developed countries.

In the United Kingdom, industrial and economic imbalance between regions is reflected in, among other things, unemployment levels. The imbalance may be due to a poor mix of industry in peripheral regions or may be due to the remoteness of peripheral regions from markets, supplies of labour, and the like which are found chiefly in core regions. The British government has at various times taken steps to redress the imbalance of industrial and economic activity between the regions. Part of the success of these steps can be seen in the relocation of manufacturing firms.

In socialist states, communist governments have inherited a pattern of regional imbalance from earlier times. Attempts to equalize the spread of economic activity in some countries, including the German Democratic Republic and Poland, have, by and large, been successful. The pattern of industry in socialist cities has also been changed to some degree in accordance with the ideals of communism.

In countries of the Third World, the differences between core regions and peripheral regions are very marked. The historical development of core and periphery is very much the same, in broad terms, in most Third World countries. In some countries, little has been done by governments to retard the growth of the core area or to stimulate the growth of the peripheral areas. In other countries, including Nigeria, some effort has been made to take industry to peripheral regions.

The location of industry in towns and cities is influenced by the same factors as industry in other areas. Although some towns and cities have a unique industrial structure, a fairly regular, broad pattern of industrial land use can be discerned in many towns and cities of the developed western world.

Further Reading

Location in Space, P. Lloyd and P. Dicken, Harper & Row (1977).
Regional Development in Britain (second edition), G. Manners *et al.*, John Wiley & Sons (1980), Chapters 1, 2, 3, and 13.
The Location of Manufacturing Industry, J. Bale, Oliver & Boyd (1977), Chapters 4, 8, and 9.
The Third World: Problems and Perspectives, edited by A. B. Mountjoy, Macmillan (1978), Chapter 2.
The Geography of Tropical African Development, A. M. O'Connor, Pergamon (1978), Chapter 5.
Latin American Development, A. Gilbert, Penguin (1974), Chapter 3.
Human Geography: Theories and their Applications, M. G. Bradford and W. A. Kent, Oxford University Press (1977), Chapters 11 and 12.
Industrial Change, edited by F. E. I. Hamilton, Longman (1978), Chapters 6 and 8.

CHAPTER THREE
THE ENVIRONMENTAL IMPACT
OF INDUSTRY

The Pollution Problem

Industrial processes are a source of many **pollutants** and many types of **pollution** (Figure 3.1). Air, soil, rivers, lakes, and oceans are all, in one place or another, polluted by industrial effluents. Towns and cities, being the places where industry tends to cluster, are commonly source areas of industrial pollutants.

Industry runs largely on power from coal, oil, and natural gas — the fossil fuels — the burning of which releases carbon dioxide, sulphur dioxide, smoke, toxic metals, and heat. Some 20 billion tonnes of **carbon dioxide** are added to the air every year by combustion of fuels. Carbon dioxide in the air absorbs some of the long-wave radiation emitted by the earth's surface. The more carbon dioxide there is, the greater the warming of the air. Prolonged emissions of carbon dioxide into the air might therefore be expected to raise average global temperatures, though the relationships are complex and the exact outcome is uncertain. Of more pressing concern to city dwellers is the fact that too much carbon dioxide in urban air may lead to chronic respiratory disorders.

Sulphur dioxide is released when fuels containing sulphur are burnt. High concentrations of sulphur dioxide irritate the lungs. Sulphur dioxide may combine with rain water to form sulphuric acid which, as well as weathering the soil at a faster than normal rate, corrodes metal and textiles and attacks buildings. Some European cities are having to spend funds on preserving rapidly weathering historic buildings and monuments. The statues on the Acropolis for instance are being disfigured by pollution from nearby Athens which houses one-third of all Greek industry. A group of plants called lichens — these include the yellow and green encrustations found on buildings and tombstones — are sensitive to sulphur dioxide concentrations in the air. They can therefore be used to delimit zones with differing levels of air pollution. Industrial furnaces also give off **nitric oxide**. Very high levels of nitric oxide have been recorded in cities but fortunately it is rapidly washed out in rain water.

Smoke, which results from the incomplete burning of fossil fuels, is a health hazard, both in its pure form and in combination with fog, that is, as smog. In the winter of 1952 to 1953, a temperature inversion over London induced the accumulation of smoke and carbon dioxide near the ground. About 4000 more people died than in a normal winter. As a result the first Clean Air Act was passed in 1956. Industry alone is not to blame for smoke pollution. In Britain, domestic, coal-burning, open fires produce 85 percent of the total smoke emissions.

Combustion and other industrial processes also release fine particles of **dust** and **grit** into the air. The larger particles may settle quickly, but the finer ones may remain suspended in the air to form mist clouds and haze which obscure sunlight and lead to less hours of bright sunshine in many British cities during winter months. Many industrial processes, along with automobiles, release **toxic metals**, such as lead and mercury, into the air. Once airborne, these metals are carried by the wind and eventually fall out or are washed out by rain to contaminate the soil.

Figure 3.1   Pollution of the air and sea by industry at Workington, Cumbria
Courtesy of Aerofilms Ltd.

Fumes from brickworks in the Midlands can lead to fluoride poisoning in cattle; effluent from the Fort William aluminium smelter has contaminated nearby pastures with fluorine; lead-smelting has in a number of places caused the death of farm stock. But the broadcasting of toxic metals does not take place over short distances only: the level of lead concentration measured in deep snow samples from the Greenland ice cap has increased a thousandfold over the last two centuries.

Industrial processes generate **heat.** Part of this heat may contribute to the urban heat island (see *Settlements*). More locally, heat from cooling systems, especially those of power stations, may cause a rise of up to 10 °C in stream and sea temperatures in the vicinity of the outlet. This thermal pollution can cause an imbalance in aquatic life but the effect appears to be small.

The pollution of streams, lakes, and seas by the dumping of municipal, industrial, and agricultural **wastes** is widespread. Many British rivers are polluted for some distance downstream of cities. Wastes dumped in Lake Ontario have boosted the nutrient concentrations in the lake water; this has led, through a series of ecological changes which involve a vast increase in the growth of algae, to a drastic decline in catches of commercially valuable fish, by a factor of a hundred or a thousand for most species. Organic waste dumped in the Baltic Sea leads to oxygen being used up by bacteria who decompose the waste. Many parts of the Baltic are now very deficient in oxygen and some areas are devoid of it and cannot support life. The dumping of wastes in the sea is carried out hand in hand with the dumping of wastes on the land. Spoil heaps and refuse tips are unattractive and of little value to agriculture because they are low in nutrients and, at many sites, contain toxic metals. A danger at chemical dumps is that poisonous chemicals may leak slowly into ground water.

The Pollution of Air

Many problems of air pollution stem from the inadequate dispersion of pollutants by wind. **Air stability** is the key to understanding how airborne pollutants can become concentrated at ground level. Of particular importance is the development of **temperature inversions**, layers of warm air over layers of cool air. To see the link between pollution, air stability, and temperature inversions, a little background work on stable and unstable air will be helpful.

STABLE AND UNSTABLE AIR

The temperature of air tends to decrease with height above the ground. Why should this be so? A parcel of air, if forced to rise, encounters a more and more rarefied atmosphere; it therefore expands. In expanding, the parcel of air uses up energy. Where does this energy come from? It cannot come from neighbouring parcels of air for they too are expanding. The answer is it comes from within the parcel itself, in fact from the heat energy stored in the parcel. So the parcel expands by using up some of its heat reserves. The result is that the expansion leads to the cooling of the parcel of air.

If the parcel of air is dry or unsaturated (more water vapour can be added to it before condensation or the formation of water droplets takes place), the rate of cooling is constant with height. This, the **dry adiabatic lapse rate**, is 10 °C for every kilometre risen, which is the same as 1 °C for every hectometre (hundred metres) risen. The word adiabatic signifies that no source of energy external to the parcel of air has been involved in the cooling process. Should a parcel of air be forced to sink, adiabatic warming will take place at 1 °C for every hectometre of descent. The warming takes place because, on descending, the air parcel meets more and more dense air and is compressed, the compression leading to a rise in temperature.

If a parcel of air cools to the temperature at which water condenses to form water droplets by the process of condensation (this is called the dew-point temperature), the lapse rate changes. Condensation releases latent heat (heat taken up by water vapour when evaporated). This heat partly offsets the heat loss due to adiabatic cooling. The **saturated adiabatic lapse rate**, as it is known, is thus less than the dry adiabatic lapse rate at roughly 6 °C per kilometre, though it is not constant. If a saturated parcel of air is forced to sink it will warm at the saturated adiabatic lapse rate.

The dry and saturated adiabatic lapse rates describe temperature changes in a parcel of air which is forced to rise or to sink. The actual variation of air temperature with height, as might be measured by an ascending balloon carrying equipment to record and transmit temperature data to a receiving station, is known as the **environmental lapse rate**. By comparing the environmental lapse rate with the dry and saturated adiabatic lapse rates, a picture of air stability at different heights can be built up. Consider the following cases.

Case one   The ground temperature is 20 °C. The environmental lapse rate is 0.9 °C per hectometre; this is less than the dry adiabatic lapse rate. Under these conditions, a small parcel of air forced to rise by, say, a turbulent eddy at the ground will cool adiabatically (Table 3.1). This cooling keeps the parcel cooler (and so heavier) than its surroundings (Table 3.1), and the parcel will sink back to ground level. The air is thus stable and will not be vulnerable to turbulent eddies and other disturbances.

Case two   The ground temperature is again 20 °C. The environmental lapse rate is 1.1 °C per hectometre, which is more than the dry adiabatic lapse rate. A parcel of air forced to rise from the ground will cool adiabatically. But this cooling will not be sufficient to bring the parcel of air to the same temperature as its surroundings. The parcel of air will always be warmer (and so lighter) than its surroundings and, having been set in motion, it will tend to keep rising. The air is thus unstable.

1 Plot the temperature profiles for cases one and two, using the data in Table 3.1, on Figure 3.2. The temperature profile which would result from the adiabatic cooling of a parcel of air is drawn for you.

2 Assume that the dew-point temperature for the air in case one is 16 °C. Once the parcel of air has cooled to this temperature, which it does at 4 hectometres, condensation starts and the cooling of the parcel drops to the saturated adiabatic lapse rate of 0.6 °C per hectometre.

Complete Table 3.2 and plot the results on Figures 3.2. Mark on Figure 3.2 the regions of stable and unstable air above 4 hectometres, and mark the condensation level, the level at which dew-point temperature is reached.

Table 3.1   Temperature (°C) changes with height

| | Height above ground (hm) | | | | | Stability of air |
Lapse rates	0	1	2	3	4	
Case one (Environmental lapse rate 0.9 °C per hm)	20.0	19.1	18.2	17.3	16.4	Stable
Dry adiabatic lapse rate (1.0 °C per hm)	20.0	19.0	18.0	17.0	16.0	—
Case two (Environmental lapse rate 1.1 °C per hm)	20.0	18.9	17.8	16.7	15.6	Unstable

INVERSIONS AND POLLUTION

In unstable air, turbulent eddies thoroughly mix pollutants with the air. If the air is unstable up to at least a kilometre or so above ground level, the mixing takes place in a large enough volume of air to keep pollution concentrations low: the dispersion of pollutants

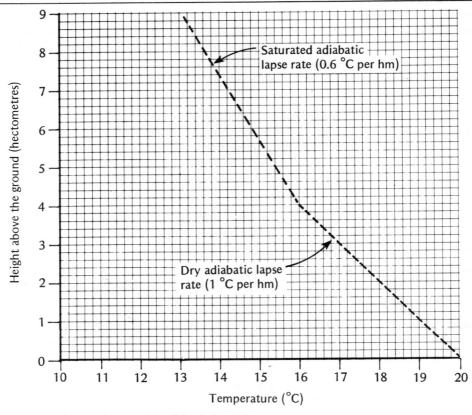

Figure 3.2   Dry and saturated adiabatic lapse rates

Table 3.2   Temperature changes (°C) when parcel is saturated

| | Height above ground (hm) | | | | | Stability of air |
Lapse rates	4	5	6	7	8	
Case one (Environmental lapse rate 0.9 °C per hm)	16.4	15.5				
Saturated adiabatic lapse rate (0.6 °C per hm)	16.0	15.4	14.8	14.2	13.6	—

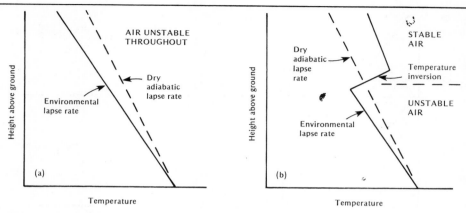

Figure 3.3   Explaining a temperature inversion
   (a) The environmental lapse rate is less than dry adiabatic lapse rate. Air is unstable at all heights, turbulence and convection are not restricted, and pollutants will disperse.
   (b) A temperature inversion leads to a layer of stable air above a layer of stable air. Turbulence and convection are restricted by the bottom of the stable layer and concentrations of pollutants in the unstable layer near the ground increase. The base of the inversion is sometimes called the convective lid.

is normally effective (see Figure 3.3a). Such conditions occur on sunny days, especially in summer. In stable air, turbulent eddies are suppressed and little mixing takes place; pollutants become concentrated near the ground. A layer of stable air overlying a layer of unstable air will act as a lid which prevents the pollutants dispersing upwards, even though the air is unstable near the ground (Figure 3.3b).

Stable air is commonly associated with a **temperature inversion**. An inversion is a layer of warm air lying over a layer of cool air (Figure 3.3b).

1   Study Figure 3.4 which shows typical plume patterns from stacks under various conditions of air stability.

   a   Label the regions of stable and unstable air on the temperature profiles at the left of the figure.

   b   Try to explain the plume patterns in terms of stability. For instance, the looping pattern is produced in unstable air, the plume following the tortuous path taken by turbulent eddies. (Though it was stated earlier that pollutants are dispersed in unstable air, looping can bring a thick cloud of pollutants to the ground, but such high pollutant concentrations will occur only locally.)

   c   How do you think stack height might be important in alleviating problems of pollution at ground level, especially in areas prone to temperature inversions?

Temperature inversions may be brought about by cooling from below, warming from above, and advection of warmer or cooler air. We shall now describe some pollution problems associated with each of these processes.

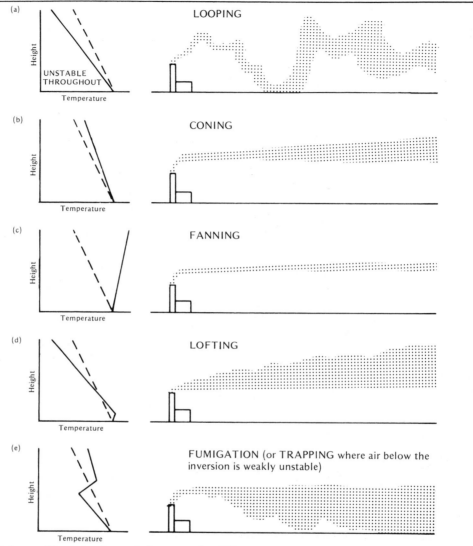

Figure 3.4 The behaviour of plumes under different conditions of air stability. In all figures, the dotted line is the dry adiabatic lapse rate and the solid line is the environmental lapse rate. Adapted from 'Some restrictive meteorological conditions to be considered in the design of stacks' by E. W. Bierly and E. W. Hewson (1962), *Journal of Applied Meteorology*, 1, 383–390

The Disaster at Donoro

The Pennsylvanian town of Donoro lies in a sheltered valley. In October 1948 it was experiencing anticyclonic weather with clear skies and light winds. At night, the ground cooled fast because under the clear skies much long-wave radiation was lost to the atmosphere but little was received from clouds. The cold ground chilled the lowermost layers of air and a radiation inversion developed. This inversion was made stronger by cold air draining down valley sides and collecting in the valley bottom, where Donoro lay, as a lake of cold air. Normally, sunlight would warm up the cool air during the day and so get rid of the inversion. Because the valley was sheltered from the sun by mountains, which in autumn was at a low angle, the inversion remained, capping the cold air lake which continued to grow. After a few days the inversion level was above the tallest stacks in the town. Factories continued pouring out pollutants and, trapped under the stable air at the inversion base, noxious chemicals filled the air. This situation lasted six days, during which time several thousands of people became ill and twenty died through breathing the highly polluted air.

2 How could the disaster have been averted?

Smog in Los Angeles

In anticyclones (areas of high atmospheric pressure) air commonly sinks gently between about 0.5 and 5 kilometres above the earth's surface. In sinking, the air encounters greater pressure, is compressed, and so warms. If the air is unsaturated it will warm at the dry adiabatic lapse rate. The result of the warming is a bulge in the temperature profile; this forms a subsidence inversion creating a lid of warm, stable air above cool, possibly unstable air over large areas. Anticyclones not infrequently stagnate over a region for a couple of weeks, so subsidence inversions can be persistent features. A large high pressure system which lies over the Pacific Ocean frequently extends across to the mountains which girdle the Los Angeles basin. Sinking air in this high pressure system warms to form a subsidence inversion at between 300 and 900 metres above Los Angeles. During the night a radiation inversion may also develop near the ground. Shortly after dawn, as factories and power stations start up and cars pour onto the roads, pollutants begin to fill the air. But trapped by the stable air in the two inversions the pollutants cannot be dispersed — the city becomes smothered in an eye-watering, murky smog.

3 How could the smog problem in Los Angeles be solved?

Pollution in an Urban Plume

Air in rural areas commonly develops a radiation inversion at night (Figure 3.5). In drifting over a relatively warmer city, the lower part of the inversion is warmed, the inversion base lifts to be replaced by unstable air. The further across the city the air drifts, the higher the inversion base

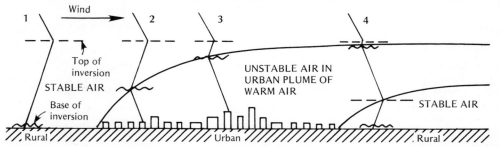

Figure 3.5   An elevated inversion due to stable rural air moving over a warm city at night. 1, 2, 3, and 4 are temperature profiles.
Reprinted with permission from *Boundary Layer Climates* by T. R. Oke (1978), published by Methuen, London, figure 9.3e

lifts (Figure 3.5). Once downwind of the city, chilling of the lowermost layers of air by the cold rural ground reestablishes the radiation inversion, though aloft unstable air of the urban plume (Figure 3.5) is found surmounted by the last vestiges of the elevated inversion produced on the upwind side of the city (Figure 3.5).

4   Figure 3.6 shows plume behaviour from stacks in and around a city at night under clear skies and light winds. Explain the pattern of plume behaviour by studying the air stability conditions delimited on Figure 3.5.

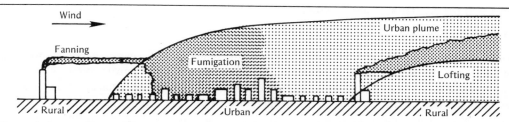

Figure 3.6   Plume behaviour in a city at night with clear skies and light winds
Reprinted with permission from *Boundary Layer Climates* by T. R. Oke (1978), published by Methuen, London, figure 9.10b

SOURCES OF POLLUTION

The more industry a region contains, the greater its energy consumption.

1   To what extent is the above statement borne out by the information conveyed in Figure 3.7?

2   We should expect regions with the highest energy consumptions to have the highest levels of pollution.

  a Plot the data listed in columns 2 and 3 of Table 3.3 as a graph (Figure 3.8a).

  b Comment on the relationship between sulphur dioxide concentration and energy consumption per unit area.

  c Plot the data listed in columns 2 and 4 of Table 3.3 as a graph (Figure 3.8b).

  d Bearing in mind the information given in Table 3.3, column 5, comment on the relationship between mean smoke concentration and energy consumption per unit area.

  e Using an atlas, explain the pattern of sulphur dioxide emissions over Europe as mapped in Figure 3.9.

3   Study Table 3.4 which shows the amount of grit and dust deposition in different environments in England and Wales. List possible sources of dust and grit in the air, then try to explain the differences brought out in Table 3.4.

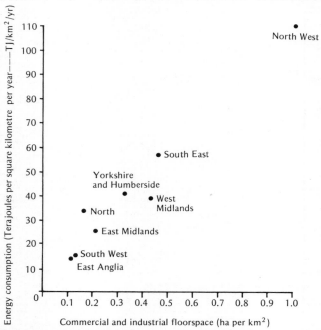

Figure 3.7 The relationship between energy consumption and industrial and commercial floorspace by regions in England

Table 3.3 Some pollution data for English regions

1	2	3	4	5
Region	Energy consumption (TJ/km^2/yr)	Mean concentration of sulphur dioxide at individual urban sites (μg/m^3)	Mean smoke concentration at individual urban sites (g/m^3)	Percentage of total domestic coal sales for naturally smokeless fuels
North	34.06	59	37	2.1
Yorkshire and Humberside	41.11	88	38	0.9
East Midlands	25.65	62	35	5.5
East Anglia	14.41	52	25	35.3
South East*	57.45	79	24	36.2
South West	15.85	44	19	27.9
West Midlands	39.01	76	37	3.4
North West	110.97	90	39	4.3

*Mean values for Greater London and South East excluding Greater London
Source of data: Department of the Environment, *Digest of Environmental Statistics* (1976)

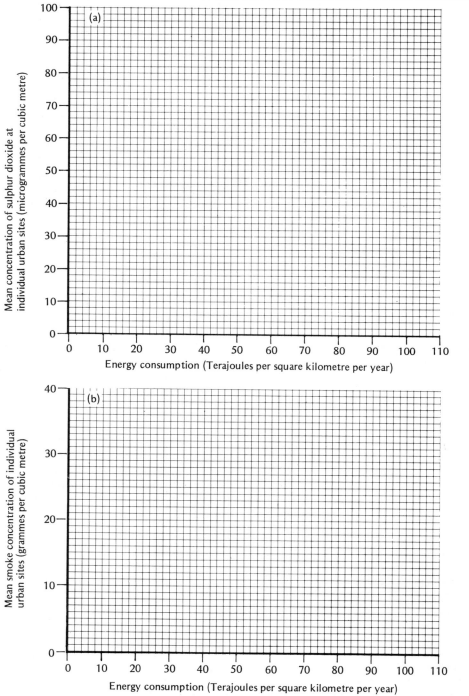

Figure 3.8

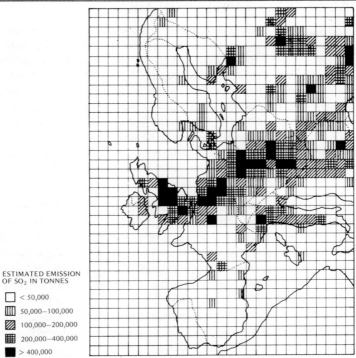

ESTIMATED EMISSION
OF SO₂ IN TONNES

☐ < 50,000

▥ 50,000–100,000

▨ 100,000–200,000

▦ 200,000–400,000

■ > 400,000

Figure 3.9   Estimated emission of sulphur dioxide in Europe in 1973
Reprinted with permission from 'Monitoring long-range transport of air pollutants: the OECD study' by B. Ottar (1976), *Ambio*, V(5–6), 203–206, figure 2

Table 3.4   Grit and dust in various environments

Environment	Grit and dust deposition rate 1973/1974 (Median values from a number of sites in each environment measured as milligrams per square metre per day)
Purely industrial area	148
An area of dense, old-fashioned housing	83
An area of low-density housing	77
The commercial centre of a town	99
Open ground in the centre of a town	73
The outskirts of a town	66
In open countryside	53

Source of data: Department of the Environment, *Digest of Environmental Statistics* (1976), table 1.10

THE REMOVAL OF POLLUTANTS FROM AIR

The saying 'what goes up must come down' is true of pollutants emitted into the atmosphere. Air pollutants are removed from the air by three chief processes. The first process involves the **settling out** of relatively large particles of grit and grime (say, larger than 1 micrometre in diameter) under the influence of gravity. Even strong turbulent eddies are unable to hold particles larger than 10 micrometres in diameter in the air for long and they fall to earth near the place they were emitted. Particles between 1 and 10 micrometres in diameter remain airborne a little longer and so travel farther. In theory, particles less than one micrometre in diameter can remain suspended in the atmosphere indefinitely; in practice, they may join to form larger aggregates which will settle out, or may be removed by one of the two other processes responsible for ridding the air of pollutants, namely, dry deposition and scavenging by precipitation.

Dry deposition is the adsorption of pollutants by the land surface when the pollutants are carried along by a turbulent wind. **Scavenging** by precipitation, or **wet deposition**, is usually the most effective process for clearing the atmosphere of pollutants in the form of gases or small particles. Two mechanisms are involved. Some small particles and gases are delivered to clouds by convective updraughts of air. Once in a cloud (or fog), the particles may act as centres or nuclei around which condensation of water or deposition of ice may take place. If the water droplet or ice particle grows big enough, it will fall to the ground carrying the pollutant with it; this mechanism is called **rainout**. Below a cloud, falling precipitation may sweep up pollutants in its path and then carry them to the ground; this mechanism is called **washout** and is far more efficient a cleansing process than rainout.

1 Study Figures 3.10a and b which are maps of dry and wet deposition of sulphur dioxide in Europe for the period December 1973 to March 1975. Bearing in mind the source of sulphur dioxide emissions (Figure 3.9) and the pattern of rainfall (consult an atlas), explain the pattern shown.

POLLUTION IN GREATER MANCHESTER

An interesting study of urban pollution has been made in Manchester. Greater Manchester is a large industrial region with an area of about 520 square miles and a population nearing three million; it harbours many pollution-generating activities, the chief among which is the burning of fuel by industry, automobiles, and houses to produce heat and power. Important sources of pollution — industrial plants and electricity-generating stations — are shown in Figure 3.11a.

1 Compare the sources of pollution in Manchester with the pattern of sulphur dioxide concentration for 1963 to 1964 (Figure 3.11b).

In the Greater Manchester study, a **composite pollution index** was derived; this, as well as the pollution of the urban air, took into account the pollution of urban land and water, and noise pollution. The index, which ranges from zero for a pollution-free area to 100 for an area suffering greatest pollution, was computed from data on smoke concentraton, sulphur dioxide

Figure 3.10   (a) Estimated sulphur dioxide dry deposition pattern (g/m²) for the period December 1973 to March 1975
(b) Estimated wet deposition pattern for sulphate (g/m²) for the period December 1973 to March 1975

(a) and (b) reprinted with permission from 'Monitoring long-range transport of air pollutants: the OECD study' by B. Ottar (1976), *Ambio*, V(5–6), 203–206, figures 5 and 6

concentration, traffic density, the area of spoil heaps and the like, the level of oxygen demand in rivers (see p. 171), and the level of oxygen demand in canals. The composite pollution index varies throughout the region (Figure 3.11c).

2   To what extent does the pattern of the composite pollution index reflect the socio-economic index, which measures the occupational structure in the region, a low value indicating high occupation status, shown in Figure 3.11d?

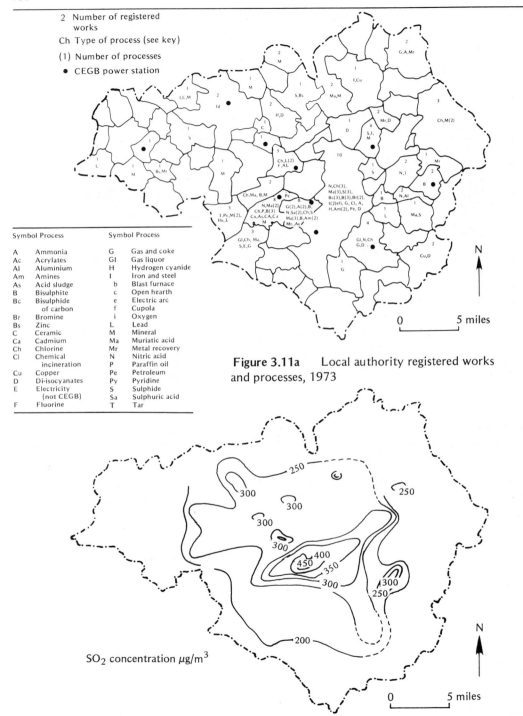

2 Number of registered works
Ch Type of process (see key)
(1) Number of processes
• CEGB power station

Symbol	Process	Symbol	Process
A	Ammonia	G	Gas and coke
Ac	Acrylates	Gl	Gas liquor
Al	Aluminium	H	Hydrogen cyanide
Am	Amines	I	Iron and steel
As	Acid sludge	b	Blast furnace
B	Bisulphite	c	Open hearth
Bc	Bisulphide of carbon	e	Electric arc
Br	Bromine	f	Cupola
Bs	Zinc	i	Oxygen
C	Ceramic	L	Lead
Ca	Cadmium	M	Mineral
Ch	Chlorine	Ma	Muriatic acid
Cl	Chemical incineration	Mr	Metal recovery
Cu	Copper	N	Nitric acid
D	Di-isocyanates	P	Paraffin oil
E	Electricity (not CEGB)	Pe	Petroleum
F	Fluorine	Py	Pyridine
		S	Sulphide
		Sa	Sulphuric acid
		T	Tar

Figure 3.11a   Local authority registered works and processes, 1973

SO_2 concentration $\mu g/m^3$

Figure 3.11b   Greater Manchester mean sulphur dioxide concentration contours, winter 1963/1964

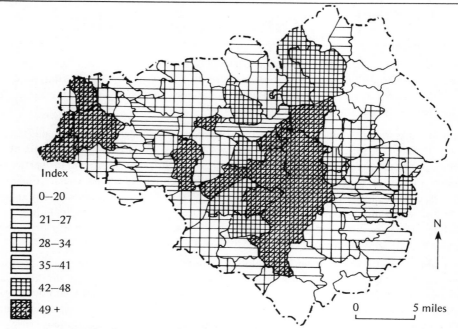

Figure 3.11c   Local authority composite pollution indices

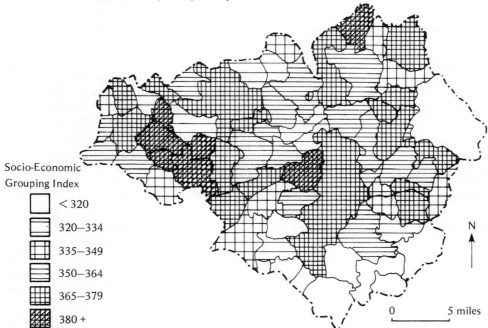

Figure 3.11d   Local authority socio-economic grouping indices, 1966
Figures 3.11a, b, c, and d are reprinted with permission from *The Geography of Pollution: A Study of Greater Manchester* by C. M. Wood *et al.* (1974), published by Manchester University Press, figures 6, 11, 15, and 31

The Pollution of Water

Lakes, rivers, and coastal waters of all industrialized countries are polluted, often severely so. The worst water pollution is concentrated where waste water from towns and from industry is fed directly into rivers, lakes, or the sea. But no water body anywhere around the globe, not even the deep waters of the oceans, is pollution free.

Water pollutants include domestic sewage, inorganic wastes from agriculture and industry, radioactive materials, solid objects like old bicycles and cars, and dredging spoil. Some of these pollutants act directly to change the physical and chemical make-up and properties of the water — its saltiness, acidity, oxygen level, and so on. Other pollutants, usually the more dangerous, disturb physiological processes in aquatic life, including reproduction. Though they may be discharged into water bodies at low concentrations, toxic substances, including radioactive ones, pass along food chains and by the time they reach the top carnivore level they may be 100, 000 times more concentrated (see *Agriculture*, p. 134).

In the main, the pollution of water in developing countries is caused by insanitary conditions. For the poorer part of the population, rivers are both a latrine and a source of drinking water, a state of affairs which promotes the spread of diseases like typhoid, cholera, and dysentry. In industrialized countries, the pollution of water results mainly from the discharge of the many and various waste substances of households and industry. No matter what the cause of water pollution, the effect is to create a health hazard and reduce the recreational potential the water might have. In developing countries, lack of funds makes pollution abatement measures difficult to implement. In industrialized countries, the sheer complexity of the problem of water pollution rules out the possibility of a simple solution to it. In all countries, the problem of water pollution has arisen on a large scale since as recently as 1950. Pollution abatement schemes set up since that time have failed to reverse pollution levels, though success is in sight for some rich, industrialized countries. The outlook is depressing. Population growth and urbanization continue. Industry continues to expand. And as the world demand for food continues to rise, so bad agricultural practices which lead to quick returns have inimical effects in other parts of the environment.

NUTRIENT POLLUTION IN LAKES AND STREAMS

Nutrients applied to the land as fertilizers and nutrient-rich sewage find their way into rivers and lakes. The result is a build-up of nutrient levels in stream and lake water; this process, called **eutrophication**, can create a pollution problem. Whether or not nutrient build-up is a problem depends upon the use to which the water is to be put. For irrigation water, nutrients such as nitrogen and phosphates in abundant supply could be an advantage. Water used by industry, for cooling purposes for instance, would not be adversely affected. Contamination of drinking water is, however, a problem which cannot be solved cheaply. But the real impact of a nutrient build-up is on the balance of aquatic life.

In the natural course of development, lake water starts clear and unproductive, contains low levels of nutrients, and harbours little plant and animal life; at this stage the lake is said to be **oligotrophic** (from the Greek, meaning little-nourished). With time the lake water becomes richer in nutrients and more productive; this is the **mesotrophic** stage. Eventually, an **eutrophic** stage (from the Greek, meaning well-nourished) is reached in which the water is rich in nutrients and plant and animal life abounds. Eutrophic streams and lakes make first-rate fisheries because of their high productivity. A hectare of an oligotrophic lake yields between about 1 and 10 kilograms of fish a year. A hectare of eutrophic lake may yield up to 500 kilograms of fish a year and the yield from artificially fertilized lakes may be as high as 10,000 kilograms of fish a year.

So why is nutrient build-up a problem? The answer is that the state of balance in the aquatic community which prevails in natural eutrophic lakes may be upset when extra nutrients are put in by man. For instance, one outcome of eutrophication is the spectacular growth of blue-green algae, as a bloom on the water surface, at the expense of diatoms and green algae. Blue-green algae impart a nasty taste and smell to drinking water and clog filters. In autumn, when the dense bloom of algae dies and starts to decay, there is an enormous surge in the activity of bacteria which decompose the algae. The big bacterial population depletes the water of oxygen dissolved in it. Low oxygen levels are unable to support desirable fish like trout but allow undesirable fish like carp to proliferate. In some cases, the oxygen level is reduced so much that all fish suffocate. And in extreme cases all life, save for anaerobic (non-oxygen-requiring) bacteria, is killed.

To control eutrophication, those nutrients responsible for it must first be isolated. Nitrogen, phosphorous, and carbon have been proposed as the most likely candidates. Recent work has indicated that phosphorus is the limiting nutrient for phytoplankton (microscopic, floating water plants which include algae) production in the majority of lakes and so phosphorus is the nutrient to control. For example, Lake Erie has suffered from an overabundance of nutrients. It is a relatively shallow lake which receives a massive influx of effluents from major population centres that surround it. The type of fish living in the lake has changed: the much-prized trout and walleye have given way to perch and sunfish. A clean-up operation is under way. A large reduction in phosphorus entering the lake should greatly reduce the pollution problem.

SEWAGE POLLUTION OF RIVERS

Sewage effluent from towns may pollute rivers. The level of pollution depends on the amount and rate of sewage added and the rate at which organisms in the water can break down the sewage. In its natural state, upstream from a town, the dissolved oxygen level in clean river water supports a rich variety of aquatic plants and animals. When sewage is dumped into the river, bacteria start to decompose it and in doing so use up large quantities of dissolved oxygen. With oxygen in short supply, clean-water organisms suffer. In the decomposition zone near the point of sewage outfall, only 'undesirable' fish like carp and gar are found. In the region of the river where dissolved oxygen is reduced to a very low level — the septic zone — only organisms like fungi and sludge worms can survive. Beyond the septic zone, the water gradually returns to its clean state, dissolved oxygen levels return to normal, and clean-water plants and animals come back (Figure 3.12).

The river system of the Warwickshire Avon drains an area of 2720 square kilometres (Figure 3.13a). Most of this area is high-grade agricultural land but it includes several fast-growing towns like Coventry, Leamington, Rugby, and Redditch. Many parts of the river system show signs of pollution. Figure 3.13b portrays three pollution categories based on the level of dissolved oxygen in the river water.

1   Identify on Figure 3.13b sections of rivers which accord with the pattern of pollution depicted on Figure 3.12.

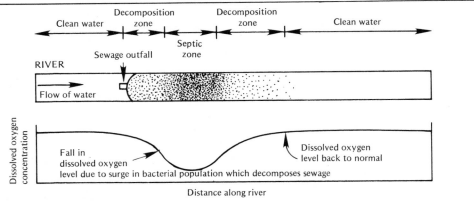

Figure 3.12   The pollution of a river by sewage

2 Study Figure 3.13c, which is a map showing the mean daily dry weather flows of the major sewage works in the Avon basin. Comparing Figure 3.13c with Figure 3.13b, identify the towns which cause bad pollution.

The stretches of river classified as badly polluted on Figure 3.13b suffer regularly from unpleasant smells, foaming, and excessive growths of algae and water weeds. But this does not much affect river users. Public water undertakings are not affected because their intakes lie upstream of major sewage outfalls. Industrial users are not inconvenienced because water of poor quality can be used as a coolant. And the water does not seem to be too polluted to be of no use for irrigating fields. Only people wishing to use the rivers for recreational purposes — boating, fishing, and the like — are hampered by the pollution.

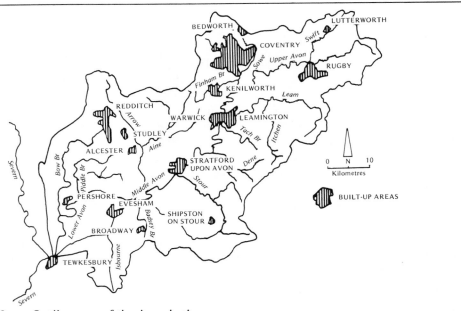

Figure 3.13a   Outline map of the Avon basin

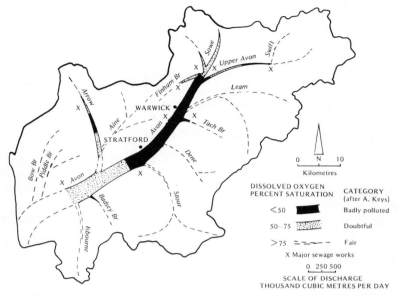

Figure 3.13b Classification of the Avon system on the basis of its median summer-period dissolved oxygen characteristics

Figures 3.13a, b, and c are reprinted with permission from 'The Warwickshire Avon: A case study of water demands and water availability in an intensively used river system' by D. C. Ledger (1972), *Transactions* of the Institute of British Geographers, 55, 83–110, figures 1, 3, and 13

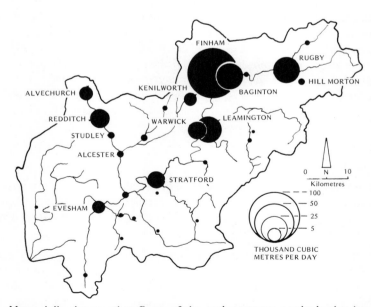

Figure 3.13c Mean daily dry weather flows of the major sewage works in the Avon basin

Derelict Land

INDUSTRY AND THE ENGLISH LANDSCAPE

An unfortunate by-product of many industrial activities is **derelict land** (Figure 3.14). The term derelict land may be taken to mean land so damaged by industrial or other development that it cannot be used for other purposes without treatment. It includes disused or abandoned land requiring restoration works to bring it into use or to improve its appearance. Land may be made derelict by mining and quarrying operations and by the dumping of wastes from towns and factories.

1   Table 3.5 gives a region-by-region breakdown of derelict land and waste disposal in England in 1974. Make choropleth maps of the data (Figure 3.15) and describe the patterns you produce.

Derelict land can be **restored**. Old quarries and pits, for instance, may sometimes be infilled with waste from mining, quarrying, commerce, industry, or households and then covered with topsoil. Between 1972 and 1974 nearly 5000 hectares of derelict land in England were restored. The quantity of waste produced each year in England is enormous. Mining and quarrying produce about 110 million tonnes, industry produces about 20 million tonnes, and households produce about 17 million tonnes. Much of this waste is used as infill for pits and excavations.

Dumped wastes can pose a problem by polluting ground water and hence the public water supply. Water percolating through freshly dumped waste is usually highly polluted, being charged with toxic metals and chlorides and nitrates in solution. After a few years, however, pollution levels become much reduced. Exceptional cases have been reported: one pit was found to be polluting ground water 80 years after dumping had ceased. If the tip lies on permeable rocks, such as sand and gravel, the percolating water will be partly filtered of pollutants, though materials in solution will still seep through to the ground water. Most tips in England lie on impermeable clay rocks and polluted water is fed into streams leading to the kind of pollution problems discussed on p. 170.

DERELICT LAND IN WALES

In Wales some of Britain's finest scenery and some of its worst scars of past industrial activity are found side by side. The most widespread and visible derelict areas are the tips of coal waste and slate debris associated with mining and quarrying. Toxic heaps of lead spoil and abandoned industrial premises also mar the landscape.

Before 1966, little had been done to tidy up these derelict areas. Between 1960 and 1966 a mere 40 hectares of derelict land were reclaimed. However, the Aberfan disaster of October 1966, in which 116 children lost their lives as a result of a flow of debris from a coal tip, brought about a change of attitude towards derelict land. During the decade following the disaster, over 3000 hectares of land were reclaimed with financial help from the government. Since 1976, the Welsh

Figure 3.14 Some by-products of industry

Development Agency has been responsible for land reclamation. Most of the reclamation has taken place in South Wales where excessive coal and iron ore extraction has left extensive areas of derelict land (Figure 3.16). Since the major role of the Development Agency is to attract new industry to Wales to reduce the high level of unemployment, it is not surprising that many land reclamation schemes involve the construction of new industrial premises. The agency's most important project has been the Hoover complex at Merthyr Tydfil. This new factory has been built on land which was formerly covered by a coal tip. The site also contained abandoned mine workings and before the plant could be built a concrete mixture had to be pumped into the underground galleries. Although this project cost £14 million, the creation of 3000 jobs is considered to make it money well spent.

Table 3.5   Derelict land and waste disposal in England, 1974

Region	Population density (persons per square kilometre)	Area of derelict land (percent of regional area)			Area covered by waste disposal (percent of regional area)	
		Spoil heaps	Extraction and pits	All forms*	Public waste	Industrial and commercial waste
North	203	0.19	0.17	0.61	0.03	0.03
Yorkshire and Humberside	317	0.08	0.08	0.35	0.04	0.09
East Midlands	239	0.07	0.07	0.33	0.03	0.02
East Anglia	142	0	0	0.14	0.02	0.007
South East	622	0.002	0.002	0.09	0.08	0.05
South West	178	0.18	0.04	0.27	0.02	0.006
West Midlands	398	0.11	0.10	0.36	0.03	0.06
North West	900	0.29	0.27	1.10	0.13	0.16
England	356	0.10	0.07	0.33	0.04	0.04

*Includes spoil heaps, extraction and pits, abandoned British Rail land, and other forms

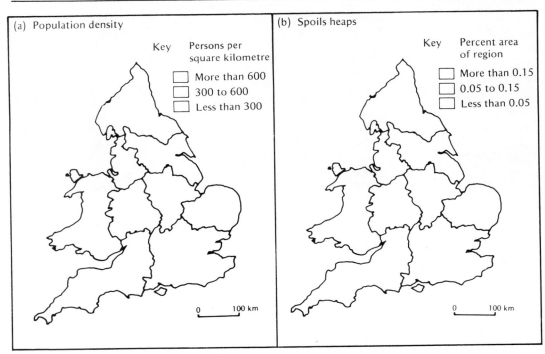

Figure 3.15   Choropleth maps of derelict land and waste disposal by region in England. The figures are the percentage of the area of each region covered by the type of land mentioned.

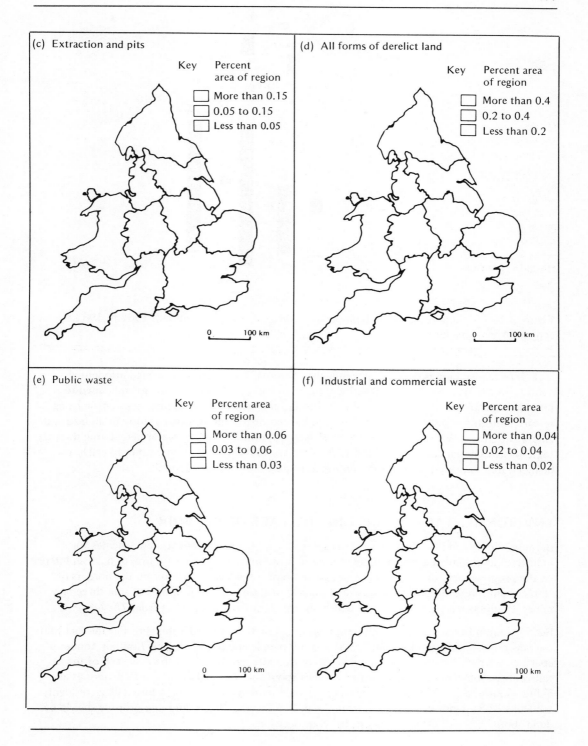

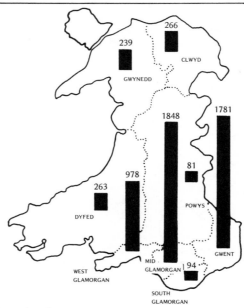

Figure 3.16   Land reclamation (completed, approved, and programmed) 1967 to 1982 by county. Area in hectares

Not all reclaimed land is used for new industry. Some of it is being transformed into leisure centres. In the Moss Valley near Wrexham, for example, coal tips have been relandscaped to provide recreational amenities such as golf, fishing, and walking. The agency is also promoting research into the vegetation of toxic land and has recently commissioned a study of all lead and zinc waste sites in Wales. In some parts of Wales, pollution of rivers by water percolating through toxic waste material has given cause for concern. Already lead waste in the Conway valley is being treated to prevent this flush pollution during periods of heavy rainfall.

ENVIRONMENTAL CONFLICT IN THE VALE OF BELVOIR

Because mining and other industrial activities may spoil the environment, new schemes for industrial development are sometimes met by opposition from people who live, or have an interest in, the proposed area of development. A case in point is the Vale of Belvoir on the borders of Nottinghamshire and Leicestershire where the National Coal Board proposes to sink three new mines. Conflicts have arisen, some of which are apparent from the extract from *The Observer*.

Bearing in mind the information given in Figures 3.17a, b, and c and Table 3.6, and the fact that the new mines would employ some 3000 to 4000 people and produce some 7 million tonnes of coal a year which would be used in power stations, write a summary of the case each of the following people might make in airing his views about the National Coal Board's proposals at a public inquiry: a spokesman for the National Coal Board; a local farmer; a local village resident; the Duke of Rutland. Bearing in mind all the pros and cons, what would you propose should be done about the conflict? Give reasons for your decision.

Digging in for the battle of Belvoir

By GEORGE BROCK

THE NATIONAL coal Board outlined last week its plans for mining the largest untapped coal reserve in Europe. It lies beneath the Vale of Belvoir in Leicestershire, and some of the 18,000 people living there want the coal to stay where it is.

The battle for Belvoir (pronounced 'Beaver') will not be joined in earnest until the summer, when the NCB decides exactly where it will want to site mineshafts. But fierce skirmishing is already under way.

The Duke of Rutland has said he will be down in front of the bulldozers and the campaigners have nearly adapted the Government 'Save It' energy slogan to their own purposes, also printing stickers reading 'Achtung — Minefield!' On a visit to a nearby colliery NCB chairman Sir Derek Ezra had to take his leave in a police car facing loud heckling.

A frisson has been felt among cheese-eaters after the revelation that mining might also threaten the future of Stilton cheese. The farmers and commuters of the Vale use words like 'cancer,' 'desecration' and 'rape' but the nearest the NCB engineers come to colourful language are names they give in their six coal seams such as Dunsil-Waterloo, Top Bright and Cinderhill Main.

Last July the NCB announced that it had found 350 million tones of coal and that it was thinking of one pithead. Last week it revised its ideas after more survey work: after 80 boreholes and 250 miles of seismic survey lines, it has settled on 500 million tons and four possible areas for sinking shafts near the villages of Hose, Langar, Asfordby and Saltby.

Preliminary noises to be heard in the broad, fertile valley suggest that the debate will be more intensely contested than the arguments over mining at Selby, the NCB's other new coalfield in Yorkshire.

The NCB starts from the position that the need for coal is established and that it would like a decision on how and where it can preceed within a year. The Vale of Belvoir Protection Group, the National Farmers' Union, and the local MP say that the need for coal is far from proved and that, even if it were, the first NCB plans were environmentally inadequate. The Stilton Cheesemakers Association is taking a gentler stance and is still discussing with the NCB the potential risks to the delicate blue-veining process.

'The nub of the thing,' said the protection group's chairman, Mr Richard Putnam, 'is not the mines, or how many their are going to be, or spoil heaps, but "do we need to mine it?"'

The NCB takes a different view: 'If we sat back and did bugger all, the mining industry would continue to contract at a rate of three or four million tons of output per year,' said Mr Peter Binns, the NCB public relations man with the unenviable job of selling the project.

When and if the Belvoir pits are in full production by the late 1980s they will produce well over three million tons a year and the NCB says that this will help to fill the predicted energy gap at the end of the century.

Reprinted by permission of *The Observer*

Table 3.6 Population of settlements in and around the Vale of Belvoir, 1971

Nottingham	300,630	Redmile and Plungar	675
West Bridgeford	28,602	Waltham-on-the-Wolds	638
Grantham	27,943	Harlaxton	623
Melton Mowbray	19,936	Eaton	551
Radcliffe-on-Trent	7,702	Holwell	543
Keyworth and Normanton	7,639	Cropwell Butler	519
Cotgrave	5,083	Langar	490
Bingham	5,053	Scalford	488
Astfordby	3,086	Woolsthorpe	488
Bottesford	1,866	Stathern	467
Horby, Hose, and Long Clawson	1,823	Sproxton	452
Whatton and Aslockton	1,599	Croxton Kerrial	435
Nether Broughton and Old Dalby	1,527	Allington	407
Cropwell Bishop	951	Belvoir	365

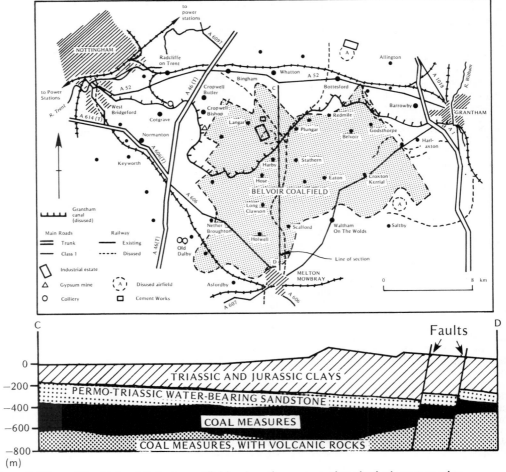

Figure 3.17a   The Vale of Belvoir coalfield — location map and geological cross-section.
The map is adapted from 'The Vale of Belvoir Coalfield: where to sink the mine?' by
D. Spooner and D. Symes (1978), *Teaching Geography*, 4(1), 3—6, figure 3

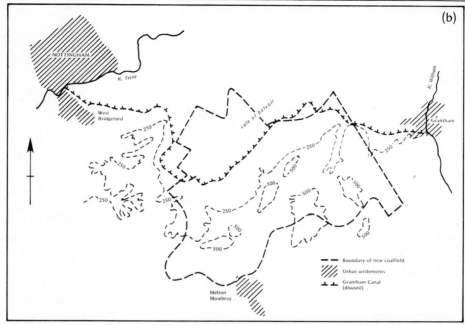

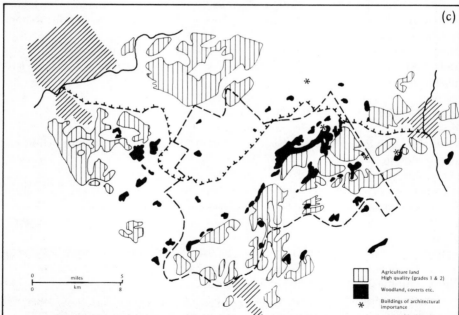

Figure 3.17   (b) Relief of the Vale of Belvoir coalfield
(c) Land quality, woodland, and buildings of architectural interest in and around the Vale of Belvoir

Reprinted with permission from 'The Vale of Belvoir coalfield: where to sink the mine?' by D. Spooner and D. Symes (1978), *Teaching Geography*, 4(1), 3–6, figures 1 and 2. © Geographical Association 1978.

GLOSSARY

Agglomeration The clustering of industries, partly as a result of external economies of scale.

Backwash effects The impoverishment of peripheral regions at the expense of a core region.

Beneficiation The process by which low-grade ores are concentrated at the site of extraction. This saves the cost of transporting bulky waste material.

Core region A thriving area which attracts, among other things, industrial investment.

Cost surface A contour map depicting place-to-place variations in the transport or production costs of an industry.

Critical isodapane A line joining points at which additional transport costs incurred in moving from the site of cheapest transport costs are just offset by the savings made using a supply of cheap labour or by agglomerating.

Cumulative causation The self-generating process by which a core region grows.

Decentralization The movement of an industry from a site in a core region to a site in a peripheral region.

Delivered price The price of a product to a customer.

Development corridor An upwards transition region lying between two cities.

Distribution costs The costs of delivering a product.

Downwards transition region A peripheral region of old, established settlements whose economic fortunes are on the wane.

Economic health The state of the economy in a region.

Economic man Someone who, in deciding where to set up a business, acts in a rational and objective way, taking all economic factors into account.

Economic recession A time of low demand and little or no investment in industry or the economy.

Entrepreneur The person or persons who take the risk of going into business. It may be an individual, the board of a private company, or the board of a state-owned company.

Expenditure All the outgoings (costs) of a business.

External economies of scale (agglomeration) The economic factors which enable some industries to reduce production costs by being located near to one another and sharing the cost of sewerage systems, insurance services, and the like.

Footloose industry An industry which has no particular market or material orientation and is free to locate almost anywhere.

Final market   The householders who buy the end-products of industry.

Friction of distance   A jargonistic phrase which refers to the fact that people and firms are more likely to move a short distance than a long distance.

Gravity model   Usually an equation which embodies the idea that the amount of movement between two places depends on some quality of the places themselves (e.g., size or attractiveness) and on the distance between them.

Growth pole   A centre of self-generating industrial and economic growth, usually in a peripheral region.

Haulage cost   The charge made for transporting goods but excluding the terminal (loading) charge.

Industrial diversity   The variety of industry in a region.

Industrial employment structure   The number of employees in different industries in a region.

Industrial estates   Areas built to house a variety of industries.

Industrial market   Industries which buy the components made by another industry.

Internal economies of scale   The savings in costs made by increasing the size of a production plant.

Isodapanes   Lines joining places with the same total transport costs.

Isotims   Lines joining places with the same procurement costs or with the same distribution costs.

Least-cost school   A group of theories of industrial location which stress the role of transport.

Linkage   The ties between allied industries (see p. 32 for a detailed example).

Location quotient   A measure of the relative importance of an industry throughout a set of regions.

Locational school   A group of people who argue that peripheral regions in the United Kingdom have problems because they are too remote from core regions.

Locational triangle   A triangle whose corners are defined by two sources of raw material and a market.

Lorenz curve   A curve used to indicate the degree of specialization of industry in a region.

Market   The place where a product is sold.

Market area   The area served by a business.

Market-area school   A group of theories of industrial location which stress the role of demand (the market).

Market orientation   The tendency of an industry to be located in or near a market.

Material index   In effect, the weight of raw materials used per tonne of product. A guide to the likely orientation of an industry.

Material orientation   The tendency of an industry to be located at or near a source of raw material.

Perfect competition   In essence, an economic climate in which no single firm can influence price and in which entry into an industry is unrestricted.

Peripheral region   A sluggish or declining area that does not usually attract industry.

Primary industry   An industry concerned with the exploitation of natural resources.

Procurement costs   The costs of buying and bringing raw materials to a production plant.

Production costs (processing costs, operating costs)   The cost of land, buildings, equipment, power (gas, oil, or electricity), and labour. In short, the costs of manufacturing and running a firm.

Profit   The difference between revenue and expenditure. A negative profit is, of course, a loss.

Pure materials   Materials which generate almost no waste during processing.

Rationalization   A process which involves a change from a lot of small production plants serving small areas to a few big production plants serving large areas.

Raw materials   The basic ingredients needed at a production plant to make the product. They may be natural materials or component parts supplied by other industries.

Regional multiplier effect   A cyclical process by which investment in a region produces side-effects which magnify the benefits accruing from the original investment.

Regional policy   A general term for measures taken by government to revitalize problem areas.

Relocation   The movement of a firm from one place to another.

Residentiary industry   An industry which tends to be found in the final market, that is, in towns and cities.

Resource frontier region   A peripheral region where new settlement is associated with the opening up of virgin territory.

Revenue   The income of a business.

Satisficer   Someone who, in deciding where to set up a business, seeks satisfactory returns (not the largest possible).

Scale of operation   The size (or capacity) and number of production plants.

Secondary industry   An industry concerned with the transformation of natural resources.

Space-cost curve   A line on a graph (or map) showing how the average cost of making a product varies from place to place.

Space-revenue curve   A line on a graph (or map) showing how the delivered price of a product varies from place to place.

Spatial margin of profitability   The bounds of an area in which revenue exceeds expenditure and production is feasible.

Spread effects   Benefits in peripheral regions resulting, in the main, from invention and innovation in a core region.

Structural school   A group of economic geographers who argue that problem regions in the United Kingdom contain the wrong mix of industry.

Substitution or opportunity cost curve   In effect, a curve depicting the trade-off between two different sorts of cost.

Terminal charge   The charge made for loading goods on to lorries or whatever.

Tertiary industry   An industry not involved in the production of material goods; a service industry.

Transport costs   The cost of hauling materials to (procuring) and from (distributing) a production plant.

Transport or freight rate   The cost of transporting a unit weight of material over a unit distance. Typical units are £ per tonne per kilometre.

Trans-shipment or break-of-bulk point   A site where the haulage of raw materials or product involves a change in mode of transport, e.g., a change from road to sea at the coast.

Upwards transition region   A peripheral region which, usually because it is well endowed with natural resources, is on the up and up.

Weight loss of gross materials   Raw materials which generate waste during processing. Zinc ore is an example.